SpringerBriefs in Statistics

T0255414

For further volumes:
http://www.springer.com/series/8921

Florentina T. Hristea
Faculty of Mathematics and Computer Science
Department of Computer Science
University of Bucharest
Bucharest
Romania

ISSN 2191-544X   ISSN 2191-5458   (electronic)
ISBN 978-3-642-33692-8   ISBN 978-3-642-33693-5   (eBook)
DOI 10.1007/978-3-642-33693-5
Springer Heidelberg New York Dordrecht London

Library of Congress Control Number: 2012949374

Printed on acid-free paper

Springer is part of Springer Science+Business Media (www.springer.com)

Florentina T. Hristea

The Naïve Bayes Model for Unsupervised Word Sense Disambiguation

Aspects Concerning Feature Selection

 Springer

To the memory of my father,
Prof. Dr. Theodor Hristea,
who has passed on to me
his love for words

Preface

The present work concentrates on the issue of feature selection for the Naïve Bayes model with application in unsupervised word sense disambiguation (WSD). It examines the process of feature selection while referring to an unsupervised corpus-based method for automatic WSD that relies on this specific statistical model. It concentrates on a distributional approach to unsupervised WSD based on monolingual corpora, with focus on the usage of the Naïve Bayes model as clustering technique.

While the Naïve Bayes model has been widely and successfully used in supervised WSD, its usage in unsupervised WSD has led to more modest disambiguation results and is less frequent. One could, in fact, say that it has been entirely dropped. The latest and most comprehensive survey[1] on WSD refers to the Naïve Bayes model strictly in conjunction with supervised WSD noting that "in spite of the independence assumption, the method compares well with other supervised methods" (Navigli 2009). It seems that the potential of this statistical model in unsupervised WSD continues to remain insufficiently explored. We feel that unsupervised WSD has not yet made full use of the Naïve Bayes model.

It is equally our belief that the Naïve Bayes model needs to be fed knowledge in order to perform well as clustering technique for unsupervised WSD. This knowledge can be fed in various ways and can be of various natures. The present work studies such knowledge of completely different types and hopes to initiate an open discussion concerning the nature of the knowledge that is best suited for the Naïve Bayes model when acting as clustering technique. Three different sources of such knowledge, which have been used only very recently in the literature (relatively to this specific clustering technique) are being examined and compared: WordNet, dependency relations, and web N-grams. This study ultimately concentrates not on WSD (which is regarded as an application) but on the issue of feeding knowledge to the Naïve Bayes model for feature selection.

---

[1] Navigli, R.: Word Sense Disambiguation: A Survey. ACM Comput. Surv. **41**(2), 1–69 (2009).

The present work represents a synthesis of 5 journal papers that have been authored or coauthored by us during the time interval 2008–2012, when our scientific interest was fully captured by the issue of feature selection for the Naïve Bayes model. This research is hereby extended, with two important additional conclusions being drawn in Chaps. 4 and 5. Each chapter will introduce knowledge of a different type, that is to be fed to the Naïve Bayes model, indicating those words (features) that should be part of the so-called "disambiguation vocabulary" when trying to decrease the number of parameters for unsupervised WSD based on this statistical model.

This work therefore places WSD with an underlying Naïve Bayes model at the border between unsupervised and knowledge-based techniques. It highlights the benefits of feeding knowledge (of various natures) to a knowledge-lean algorithm for unsupervised WSD that uses the Naïve Bayes model as clustering technique.

Our study will show that a basic, simple knowledge-lean disambiguation algorithm, hereby represented by the Naïve Bayes model, can perform quite well when provided knowledge in an appropriate way. It will equally justify our belief that the Naïve Bayes model still holds a promise for the open problem of unsupervised WSD.

Toulouse, France, November 2011   Florentina T. Hristea

Acknowledgments

The author expresses her deepest gratitude to Professor Ted Pedersen for having provided the dataset necessary for performing the presented tests and comparisons with respect to adjectives and verbs. We are equally indebted to two anonymous referees for their valuable comments and suggestions. This research was supported by the National University Research Council of Romania (the "Ideas" research program, PN II—IDEI), Contract No. 659/2009.

Contents

Acronyms

AI   Artificial Intelligence
DG   Dependency Grammar
LSA   Latent Semantic Analysis
LSI   Latent Semantic Indexing
NLP   Natural Language Processing
POS   Part of Speech
WN   WordNet
WSD Word Sense Disambiguation

Chapter 1
Preliminaries

Abstract This chapter describes the problem we are investigating and trying to solve in all other chapters. It introduces word sense disambiguation (WSD) and Naïve Bayes-based WSD, as well as local type features for unsupervised WSD with an underlying Naïve Bayes model.

Keywords Naïve Bayes model · Feature selection · Word sense disambiguation · Supervised disambiguation · Unsupervised disambiguation · Knowledge-based disambiguation · Local-type features

1.1 Introduction: The Problem

The present work concentrates on the issue of feature selection for the Naïve Bayes model, with application in unsupervised word sense disambiguation.

Word sense disambiguation (WSD), which signifies determining the meaning of a word in a specific context, is a core research problem in natural language processing, which was recognized since the beginning of the scientific interest in machine translation, and in artificial intelligence, in general. As noted in (Agirre and Edmonds 2006), finding a solution to the WSD problem is obviously essential for applications which deal with natural language understanding (message understanding, man-machine communication etc.) and is at least useful, and in some cases compulsory, for several applications which do not have natural language understanding as main goal, applications such as: information retrieval, machine translation, speech processing, text processing etc.

As a computational problem, lexical disambiguation was originally regarded as being AI-complete, that is, a problem whose solution requires a solution to complete natural language understanding or common-sense reasoning. This view originated in the fact that possible statistical approaches to the problem were almost completely ignored in the past. As it is well known, starting with the early nineties, the artificial

F. T. Hristea, *The Naïve Bayes Model for Unsupervised Word Sense Disambiguation*,
SpringerBriefs in Statistics, DOI: 10.1007/978-3-642-33693-5_1, © The Author(s) 2013

intelligence community witnesses a great revival of empirical methods, especially statistical ones. Nowadays statistical methods are used for solving a great number of problems posed by artificial intelligence, in general, and by natural language processing, in particular. WSD is one such problem, for which the Naïve Bayes model has been extensively used.[1]

In the subfield of natural language processing (from the perspective of which we shall approach WSD within the framework of the present study), the problem we are discussing here is defined as that of computationally determining which sense of a word is activated by the use of that word in a particular context and represents, essentially, *a classification problem*.

The problem becomes even more difficult to solve when taking into account the great existing number of natural languages with very high polysemy. As noted in (Agirre and Edmonds 2006), the 121 most frequent English nouns, for instance, which account for about one in five word occurrences in real English text, have on average 7.8 meanings each, according to the Princeton University lexical database WordNet (Miller 1990, 1995; Miller et al. 1990; Fellbaum 1998).

In spite of the great number of existing disambiguation algorithms, the problem of WSD remains an open one, with three main classes of WSD methods being taken into consideration by the literature: supervised disambiguation, unsupervised disambiguation and knowledge-based disambiguation.

The present study refers to unsupervised corpus-based methods for WSD. It concentrates on distributional approaches to unsupervised WSD that rely on monolingual corpora, with focus on the usage of the Naïve Bayes model as clustering technique.

Within the framework of the present study, the term "unsupervised" will refer, as in Pedersen (2006), to knowledge-lean methods, that do not rely on external knowledge sources such as machine-readable dictionaries, concept hierarchies or sense-tagged text. Due to the lack of knowledge they are confronted with, these methods do not assign meanings to words, relative to a pre-existing sense inventory, but make a distinction in meaning based on distributional similarity. While not performing a straightforward WSD, these methods achieve a discrimination among the meanings of a polysemous word.

The problem we are investigating here could be formulated in the following terms: we are given I sentences that each contain a particular polysemous word; our goal is to divide these I instances of the ambiguous word (the so-called target word) into a specified number of sense groups. These sense groups must be mapped to sense tags in order to evaluate system performance. Let us note that sense tags, as in previous studies (Pedersen and Bruce 1998; Hristea et al. 2008; Hristea 2009; Hristea and Popescu 2009), will be used only in the evaluation of the sense groups found by the unsupervised learning procedure. The discussed algorithm is automatic and unsupervised in both training and application.

From the wide range of unsupervised learning techniques that could be applied to our problem, we have chosen to use a parametric model in order to assign a sense

---

[1] Especially with reference to supervised WSD.

group to each ambiguous occurrence of the target word. As already mentioned, in each case, we shall assign the most probable group given the context as defined by the Naïve Bayes model, where the parameter estimates are formulated via unsupervised techniques. The theoretical model will be presented and its implementation will be discussed. Special attention will be paid to feature selection, the main issue of the model's implementation. Various novel methods of performing knowledge-based feature selection will be presented and discussed.

When the Naïve Bayes model is applied to supervised disambiguation, the actual words occurring in the context window are usually used as features. This type of framework generates a great number of features and, implicitly, a great number of parameters. This can dramatically decrease the model's performance since the available data is usually insufficient for the estimation of the great number of resulting parameters. A situation that becomes even more drastic in the case of unsupervised disambiguation, where parameters must be estimated in the presence of missing data (the sense labels). In order to overcome this problem, the various existing unsupervised approaches to WSD implicitly or explicitly perform a feature selection. In fact, one can say that discussions concerning the implementation of the Naïve Bayes model for supervised/unsupervised WSD focus almost entirely on the issue of feature selection.

The approach to feature selection of the present study is that of implementing a Naïve Bayes model that uses as features *the actual words* occurring in the context window[2] of the target and decreases the existing number of features by selecting a restricted number of such words, as indicated by a specific knowledge source. The size of the feature set is therefore reduced by performing knowledge-based feature selection. The Naïve Bayes model will be fed knowledge of various natures. Chapters 3, 4 and 5 will each introduce knowledge of a different type, that is to be fed to the Naïve Bayes model, indicating those words (features) which should be part of the disambiguation vocabulary when trying to decrease the number of parameters for unsupervised WSD. This type of approach will place the disambiguation process at the border between unsupervised and knowledge-based techniques, while reinforcing the benefits of combining the unsupervised approach to the WSD problem with usage of a knowledge source.

We shall ultimately compare totally different ways of feeding knowledge of various types to a knowledge-lean algorithm for unsupervised WSD based on an underlying Naïve Bayes model. The discussed method will once again prove that a basic, simple knowledge-lean disambiguation algorithm, hereby represented by the Naïve Bayes model, can perform quite well when provided knowledge in an appropriate way, a remark also made by Ponzetto and Navigli (2010).

---

[2] The context window can be of fixed size or it can be represented by the entire sentence in which the target word occurs.

1.2 Word Sense Disambiguation (WSD)

In artificial intelligence (from the perspective of which we are approaching word sense disambiguation here) the problem we are discussing is defined as that of computationally determining which sense of a word is activated by the use of that word in a specific context. In this view, word sense disambiguation (WSD) represents, essentially, a classification problem.

The importance of WSD has been widely acknowledged in recent years, with over 700 papers in the ACL Anthology mentioning the term "word sense disambiguation" and with three classes of WSD methods being taken into consideration by the literature: supervised disambiguation, unsupervised disambiguation and knowledge-based disambiguation.

Supervised disambiguation is based on learning. As it is well known, the supervised approach to WSD consists of automatically inducing classification models or rules from annotated examples. A disambiguated corpus is available for training. This disambiguated corpus will be used in training a classifier that can label words within a new, unannotated text. The task is that of conceiving a classifier which correctly classifies the new cases, based on the context where they occur. One such classifier, that has been widely used in supervised disambiguation, is the Bayes classifier, which looks at the words around an ambiguous word in a so-called context window.

Unlike supervised disambiguation, the unsupervised approach to the same problem uses no pre-existing knowledge source. Unsupervised disambiguation methods are data-driven, highly portable, robust, and offer the advantage of being language-independent. They rely either on the distributional characteristics of unannotated corpora (which will represent the approach within the present work), or on translational equivalences in word aligned parallel text. Within the framework of the present study, the term "unsupervised" will refer, as in Pedersen (2006), to knowledge-lean methods, that do not rely on external knowledge sources such as machine readable dictionaries, concept hierarchies, or sense-tagged text. Due to the lack of knowledge they are confronted with, these methods do not assign meanings to words, relative to a pre-existing sense inventory, but rather make distinctions in meaning based on distributional similarity. While not performing a straightforward WSD, these methods achieve a *discrimination among the meanings* of a polysemous word. As commented in (Agirre and Edmonds 2006), they have the potential to overcome the knowledge acquisition bottleneck (manual sense-tagging).

Unsupervised disambiguation is considered extremely important because it uses no pre-existing knowledge source (which makes it very applicable) and because it is language-independent. Even though, for both these reasons, its performance is 5–10 % lower than that of some of the dictionary-based algorithms, it also offers the advantage that it can be easily adapted to produce distinctions between usage types that are more fine-grained than would be found in a dictionary. (For example, it can distinguish between *civil suit* and *criminal suit*, while regular dictionaries record only *law suit*.) Usually, the induced clusters do not line up well with dictionary senses. If the unsupervised algorithm is run for a large number of senses, then it will split dictionary senses into fine-grained contextual variants. Information retrieval is an application for which this is considered useful.

Finally, knowledge-based disambiguation methods perform sense disambiguation (and not sense discrimination) by means of a pre-existing sense inventory. These methods can usually be applied to all words of a given text, unlike the techniques based on corpora, which can be used only in the case of those words for which annotated corpora are available.

With the exception of the case when it is unsupervised, the problem of WSD requires establishing a sense inventory, namely determining all meanings which can be assigned to each word that must be disambiguated. However, the concept of word sense still generates debates among linguists. That is probably why, nowadays, an official and unique sense inventory for English still doesn't exist. Some of the most frequent sources used for establishing a sense inventory are: electronic dictionaries, thesauri (LDOCE, Roget's Thesaurus), bilingual dictionaries in electronic format, and lexical knowledge bases (of type WordNet). Princeton University's WordNet[3] (Miller 1990, 1995; Miller et al. 1990; Fellbaum 1998) has probably become the most widely used source for establishing a sense inventory. We shall be making use of it in Chap. 3.

1.3 Naïve Bayes-Based WSD at the Border Between Unsupervised and Knowledge-Based Techniques

Unlike previous approaches (Pedersen and Bruce 1998) that, when implementing the Naïve Bayes model, make use of a small number of local features, the present work means to implement a Naïve Bayes model that uses as features the actual words occurring in the context window of the (ambiguous) target. Our study implements the model in its simplest and most straightforward form, while selecting a restricted number of words in order to decrease the number of features used and, as a result, to increase the performance of the disambiguation process. The method of performing feature selection will place the disambiguation process at the border between unsupervised and knowledge based techniques. The obtained disambiguation methods and corresponding results, as compared to previously existing ones, will reinforce the benefits of combining the unsupervised approach to the WSD problem with usage of a knowledge source for feature selection. Various types of knowledge sources (which have led to different ways of performing feature selection) will be examined. In fact, as noted in Sect. 1.1, one can say that discussions concerning the implementation of the Naïve Bayes model for supervised/unsupervised WSD focus almost entirely on the issue of feature selection.

Two early approaches to word sense discrimination, context group discrimination (Schütze 1998) and McQuitty's Similarity Analysis (Pedersen and Bruce 1997, 1998), rely on totally different sets of features and still represent the main approaches to feature selection.

---

[3] Available at http://wordnet.princeton.edu/.

As commented in (Pedersen 2006), Schütze (1998) represents contexts in a high dimensional feature space that is created using a separate large corpus (referred to as the training corpus). While Schütze (1998) reduces dimensions by means of LSI/LSA, Pedersen and Bruce (1997) define features over a small contextual window (local context) and select them to produce low dimensional event spaces. They make use of a small number of first-order features to create matrices that show the pairwise (dis)similarity between contexts. They rely on local features that include co-occurrence and part of speech information near the target word. Three different feature sets, consisting of various combinations of features of the mentioned types, were defined in (Pedersen and Bruce 1998) for each word and were used to formulate a Naïve Bayes model describing the distribution of sense groups of that word. Unlike Schütze (1998), Pedersen and Bruce (1998) select features from the same test data that is being discriminated, which, as noted in (Pedersen 2006), is a common practice in clustering in general.

The more recent disambiguation results obtained when using knowledge-based feature selection were compared to those of Pedersen and Bruce (1998) since all these disambiguation methods (see Chaps. 3, 4 and 5) use an algorithm of the same type i.e. unsupervised and based on an underlying Naïve Bayes model. Moreover, the model parameters are estimated in the same way, namely by means of the EM algorithm.

The entire discussion that is to follow concerns feature selection for the Naïve Bayes model and will be studying various types of features which differ completely from the initially and only ones used for this type of problem (Pedersen and Bruce 1998).

1.3.1 Pedersen and Bruce Local-Type Features

When performing unsupervised word sense disambiguation with an underlying Naïve Bayes model, Pedersen and Bruce (1997, 1998), define three different feature sets for each word and use them to formulate such a model describing the distribution of sense groups of that word. The feature sets taken into account were composed of various combinations of the following five types of features:

Morphology The feature denoted M represents the morphology of the ambiguous word. In the case of nouns, for instance, M is binary indicating singular or plural. For verbs, the value of M indicates the tense of the verb and can have up to seven possible values. This feature is not used for adjectives.

Part-of-speech The features denoted PL_i and PR_i represent the part-of-speech (POS) of the word i positions to the left or right, respectively, of the ambiguous word. Each POS feature can have one of five possible values: noun, verb, adjective, adverb or other.

Co-occurrences The features denoted C_i are binary variables representing whether the ith most frequent content word in all sentences containing the ambiguous word occurs anywhere in the sentence being processed.

Unrestricted collocations The features denoted UL_i and UR_i are features with 20 possible values that indicate if one of the top 19 most frequent words occurs in position i to the left (UL_i) or right (UR_i) of the target word.

Content collocations The features denoted CL_1 and CR_1 indicate the content word occurring in the position 1 place to the left or right, respectively, of the ambiguous word. In general, features (CL_i, CR_i) are identical to the unrestricted collocations, except they exclude function words and only represent content words.

All these features[4] are defined over a small contextual window (local-context) and are selected to produce low dimensional event spaces.

The three feature sets used in the experiments presented in (Pedersen and Bruce 1998) were designated A, B and C and were formulated as follows:

$$A: M, PL_2, PL_1, PR_1, PR_2, C_1, C_2, C_3$$
$$B: M, UL_2, UL_1, UR_1, UR_2$$
$$C: M, PL_2, PL_1, PR_1, PR_2, CL_1, CR_1$$

It is our belief that the most interesting aspect of the described approach is represented by the choice of such types of features and feature sets in order to formulate a Naïve Bayes model. However, as the authors note in (Pedersen and Bruce 1998) "while frequency-based features, such as those used in this work, reduce sparsity, they are less likely to be useful in distinguishing among minority senses".

Pedersen and Bruce consider nouns, verbs and adjectives as possible target words in the discrimination task, and explore the use of several different combinations of features. The two mentioned authors conducted an experimental evaluation in (Pedersen and Bruce 1998) relative to the 12-word sense-tagged corpus of Bruce et al. (1996) as well as with the *line* corpus (Leacock et al. 1993).

The obtained performance when using the described type of local-context features is relatively low. The best results were obtained in the case of nouns, where in combination with a specific feature set the obtained accuracy improved upon the most frequent sense by at least 10%. The most modest results (accuracy) were obtained in the case of the noun *line*.

No feature set resulted in greater accuracy than the most frequent sense for verbs and adjectives. In the case of nouns McQuitty's method performed better. In combination with feature set **B** it improved upon the most frequent sense by at least 10%. Pedersen and Bruce (1998) found that feature set **B** performs best for nouns, while feature set **C** performs best for both adjectives and verbs. Their disambiguation results were compared to those reported in (Hristea et al. 2008; Hristea 2009; Hristea and Popescu 2009) where an algorithm of the same type (unsupervised with an underlying Naïve Bayes model) is examined and where knowledge-based feature selection is performed. In the case of all parts of speech, test results have shown that feature selection using a knowledge source of type WordNet is more effective in sense disambiguation

---

[4] For more details concerning these types of features, feature sets, and their usage, see (Pedersen and Bruce 1998).

than local-type features are (see Chap. 3). Further comparisons were made (Preoţiuc and Hristea 2012) with web N-gram feature selection (for which see Chap. 5).

In the chapters that follow we shall be presenting three other, completely different, ways of performing feature selection for the Naïve Bayes model, when acting as clustering technique in unsupervised WSD. The full presentation of these feature selection methods will once again reinforce the benefits of combining the unsupervised approach to the WSD problem with a knowledge source of various types. Especially since we must keep in mind that knowledge-lean methods as the one proposed in (Pedersen and Bruce 1998) can also require information that is not always available. Such knowledge-lean methods can equally have difficulties when asking for information like part of speech, for instance, especially if a part-of-speech tagger does not exist for the language under investigation.

References

Agirre, E., Edmonds, P. (eds.): Word Sense Disambiguation. Algorithms and Applications. Springer, New York (2006)

Bruce, R., Wiebe, J., Pedersen, T.: The measure of a model. In: Proceedings of the Conference on Empirical Methods in Natural Language Processing, pp. 101–112. Philadelphia, PA (1996)

Eberhardt, F., Danks, D.: Confirmation in the cognitive sciences: the problematic case of Bayesian models. Mind. Mach. **21**, 389–410 (2011)

Fellbaum, C. (ed.): WordNet: An Electronic Lexical Database. The MIT Press, Cambridge (1998)

Hristea, F.: Recent advances concerning the usage of the Naïve Bayes model in unsupervised word sense disambiguation. Int. Rev. Comput. Softw. **4**(1), 58–67 (2009)

Hristea, F., Popescu, M., Dumitrescu, M.: Performing word sense disambiguation at the border between unsupervised and knowledge-based techniques. Artif. Intell. Rev. **30**(1), 67–86 (2008)

Hristea, F., Popescu, M.: Adjective sense disambiguation at the border between unsupervised and knowledge-based techniques. Fundam. Inf. **91**(3–4), 547–562 (2009)

Leacock, C., Towell, G., Voorhees, E.: Corpus-based statistical sense resolution. In: Proceedings of the ARPA Workshop on Human Language Technology, pp. 260–265. Princeton, New Jersey (1993)

Miller, G.A.: Nouns in WordNet: a lexical inheritance system. Int. J. Lexicogr. **3**(4), 245–264 (1990)

Miller, G.A.: WordNet: a lexical database. Commun. ACM **38**(11), 39–41 (1995)

Miller, G.A., Beckwith, R., Fellbaum, C., Gross, D., Miller, K.: WordNet: an on-line lexical database. J. Lexicogr. **3**(4), 234–244 (1990)

Pedersen, T.: Unsupervised corpus-based methods for WSD. In: Agirre, E., Edmonds, P. (eds.) Word Sense Disambiguation. Algorithms and Applications. Springer, New York (2006)

Pedersen, T., Bruce, R.: Distinguishing word senses in untagged text. In: Proceedings of the Second Conference on Empirical Methods in Natural Language Processing (EMNLP-2), pp. 197–207. Providence, Rhode Island (1997)

Pedersen, T., Bruce, R.: Knowledge lean word-sense disambiguation. In: Proceedings of the 15th National Conference on Artificial Intelligence, pp. 800–805. Madison, Wisconsin (1998)

Ponzetto, S.P., Navigli, R.: Knowledge-rich word sense disambiguation rivaling supervised systems. In: Proceedings of the 48th Annual Meeting of the Association for Computational Linguistics (ACL 2010), pp. 1522–1531. ACL Press, Uppsala, Sweden (2010)

Preoţiuc-Pietro, D., Hristea, F.: Unsupervised word sense disambiguation with N-gram features. Artif. Intell. Rev. (2012). doi:10.1007/s10462-011-9306-y

Schütze, H.: Automatic word-sense discrimination. Comput. Linguist. **24**(1), 97–123 (1998)

Chapter 2
The Naïve Bayes Model in the Context of Word Sense Disambiguation

Abstract This chapter discusses the Naïve Bayes model strictly in the context of word sense disambiguation. The theoretical model is presented and its implementation is discussed. Special attention is paid to parameter estimation and to feature selection, the two main issues of the model's implementation. The EM algorithm is recommended as suitable for parameter estimation in the case of unsupervised WSD. Feature selection will be surveyed in the following chapters.

Keywords Bayesian classification · Expectation-Maximization algorithm · Naïve Bayes classifier

2.1 Introduction

The classical approach to WSD that relies on an underlying Naïve Bayes model represents an important theoretical approach in statistical language processing: Bayesian classification (Gale et al. 1992). The idea of the Bayes classifier (in the context of WSD) is that it looks at the words around an ambiguous word in a large context window. Each content word contributes potentially useful information about which sense of the ambiguous word is likely to be used with it. The classifier does no feature selection. Instead it combines the evidence from all features. The mentioned classifier (Gale et al. 1992) is an instance of a particular kind of Bayes classifier, the Naïve Bayes classifier.

Naïve Bayes is widely used due to its efficiency and its ability to combine evidence from a large number of features. It is applicable if the state of the world that we base our classification on is described as a series of attributes. In our case, we describe the context of the ambiguous word in terms of the words that occur in the context.

The Naïve Bayes assumption is that the attributes used for description are all conditionally independent, an assumption having two main consequences. The first is that all the structure and linear ordering of words within the context are ignored, leading to a so-called "bag of words model".[1] The other is that the presence of one word in the bag is independent of another, which is clearly not true in the case of natural language. However, in spite of these simplifying assumptions, as noted in (Manning and Schütze 1999), this model has been proven to be quite effective when put into practice. This is not surprising when viewing the Bayesian model from a cognitive perspective, which is an adequate one in the case of a problem concerning natural language processing. And when taking into consideration that, as noted in (Eberhardt and Danks 2011), "without an account of the rationality of the observed input-output relation, the computational level models provide a summary of the observed data, but no rational explanation for the behaviour".

2.2 The Probability Model of the Corpus and the Bayes Classifier

In order to formalize the described model, we shall present the probability structure of the corpus $\mathscr{C}$. The following *notations* will be used: w is the word to be disambiguated (target word); $s_1, ..., s_K$ are possible senses for w; $c_1, ..., c_I$ are contexts of w in a corpus $\mathscr{C}$; $v_1, ..., v_J$ are words used as contextual features for the disambiguation of w.

Let us note that the contextual features could be some attributes (morphological, syntactical, etc.), or they could be actual "neighboring" content words of the target word. The contextual features occur in a fixed position near w, in a window of fixed length, centered or not on w. In what follows, a window of size n will denote taking into consideration n content words to the left and n content words to the right of the target word, whenever possible. The total number of words taken into consideration for disambiguation will therefore be $2n + 1$. When not enough features are available, the entire sentence in which the target word occurs will represent the context window.

The probability structure of the corpus is based on one main assumption: *the contexts $\{c_i, i\}$ in the corpus $\mathscr{C}$ are independent.* Hence, the likelihood of $\mathscr{C}$ is given by the product

$$P(\mathscr{C}) = \prod_{i=1}^{I} P(c_i)$$

Let us note that this is a quite natural assumption, as the contexts are not connected, they occur at significant lags in $\mathscr{C}$.

On considering the possible senses of each context, one gets

$$P(\mathscr{C}) = \prod_{i=1}^{I} \sum_{k=1}^{K} P(s_k) \cdot P(c_i \mid s_k)$$

---

[1] A bag is similar to a set, only it allows repetition.

A model with independent features (usually known as the Naïve Bayes model) assumes that the contextual features are conditionally independent. That is,

$$P\left(c_i \mid s_k\right) = \prod_{v_j \ in \ c_i} P\left(v_j \mid s_k\right) = \prod_{j=1}^{J} \left(P\left(v_j \mid s_k\right)\right)^{\left|v_j \ in \ c_i\right|},$$

where by $\left|v_j \ in \ c_i\right|$ we denote the number of occurrences of feature v_j in context c_i. Then, the likelihood of the corpus $\mathscr{C}$ is

$$P\left(\mathscr{C}\right) = \prod_{i=1}^{I} \sum_{k=1}^{K} P\left(s_k\right) \prod_{j=1}^{J} \left(P\left(v_j \mid s_k\right)\right)^{\left|v_j \ in \ c_i\right|}$$

The parameters of the probability model with independent features are

$$\left\{P\left(s_k\right), k = 1, ..., K \ \text{ and } \ P\left(v_j \mid s_k\right), \quad j = 1, ..., J, k = 1, ..., K\right\}$$

Notation:

- $P\left(s_k\right) = \alpha_k, k = 1, ..., K, \alpha_k \geq 0$ for all k, $\sum_{k=1}^{K} \alpha_k = 1$
- $P\left(v_j \mid s_k\right) = \theta_{kj}, k = 1, ..., K, j = 1, ..., J, \theta_{kj} \geq 0$ for all k and j, $\sum_{j=1}^{J} \theta_{kj} = 1$ for all $k = 1, ..., K$

With this notation, the likelihood of the corpus $\mathscr{C}$ can be written as

$$P\left(\mathscr{C}\right) = \prod_{i=1}^{I} \sum_{k=1}^{K} \alpha_k \prod_{j=1}^{J} \left(\theta_{kj}\right)^{\left|v_j \ in \ c_i\right|}$$

The well known Bayes classifier involves the a posteriori probabilities of the senses, calculated by the Bayes formula for a specified context c,

$$P\left(s_k \mid c\right) = \frac{P\left(s_k\right) \cdot P\left(c \mid s_k\right)}{\sum_{k=1}^{K} P\left(s_k\right) \cdot P\left(c \mid s_k\right)} = \frac{P\left(s_k\right) \cdot P\left(c \mid s_k\right)}{P\left(c\right)},$$

with the denominator independent of senses.

The Bayes classifier chooses the sense s' for which the a posteriori probability is maximal (sometimes called the Maximum A Posteriori classifier)

$$s' = \arg\max_{k=1,...,K} P\left(s_k \mid c\right)$$

Taking into account the previous Bayes formula, one can define the Bayes classifier by the equivalent formula

$$s' = \underset{k=1,...,K}{\arg\max} \left(\log P\left(s_k\right) + \log P\left(c \mid s_k\right) \right)$$

Of course, when implementing a Bayes classifier, one has to estimate the parameters first.

2.3 Parameter Estimation

Parameter estimation is performed by the Maximum Likelihood method, for the available corpus $\mathscr{C}$. That is, one has to solve the optimization problem

$$\max \left(\log P\left(\mathscr{C}\right) \mid \left\{ P\left(s_k\right), k = 1, ..., K \text{ and } P\left(v_j \mid s_k\right), \quad j = 1, ..., J, \ k = 1, ..., K \right\} \right)$$

For the Naïve Bayes model, the problem can be written as

$$\max \left(\sum_{i=1}^{I} \log \left(\sum_{k=1}^{K} \alpha_k \prod_{j=1}^{J} \left(\theta_{kj} \right)^{\left| v_j \text{ in } c_i \right|} \right) \right) \tag{2.1}$$

with the constraints

$$\sum_{k=1}^{K} \alpha_k = 1$$
$$\sum_{j=1}^{J} \theta_{kj} = 1 \qquad \text{for all } k = 1, ..., K$$

For *supervised disambiguation*, where an annotated training corpus is available, the parameters are simply estimated by the corresponding frequencies:

$$\widehat{\theta}_{kj} = \frac{\left| \text{occurrences of } v_j \text{ in a context of sense } s_k \right|}{\displaystyle\sum_{j=1}^{J} \left| \text{occurrences of } v_j \text{ in a context of sense } s_k \right|},$$

$$k = 1, ..., K; \ j = 1, ..., J$$

$$\widehat{\alpha}_k = \frac{\left| \text{occurrences of sense } s_k \text{ in } \mathscr{C} \right|}{\left| \text{occurrences of } w \text{ in } \mathscr{C} \right|}, \quad k = 1, ..., K$$

For *unsupervised disambiguation*, where no annotated training corpus is available, the maximum likelihood estimates of the parameters are constructed by means of the Expectation-Maximization (EM) algorithm.

For the unsupervised case, the optimization problem (2.1) can be solved only by iterative methods. The Expectation-Maximization algorithm (Dempster et al. 1977) is a very successful iterative method, known as very well fitted for models with missing data.

Each iteration of the algorithm involves two steps:

- estimation of the missing data by the conditional expectation method (E-step)
- estimation of the parameters by maximization of the likelihood function for complete data (M-step)

The E-step calculates the conditional expectations given the current parameter values, and the M-step produces new, more precise parameter values. The two steps alternate until the parameter estimates in iteration $r + 1$ and r differ by less than a threshold ε.

The EM algorithm is guaranteed to increase the likelihood $\log P(\mathscr{C})$ in each step. Therefore, two stopping criteria for the algorithm could be considered: (1) Stop when the likelihood $\log P(\mathscr{C})$ is no longer increasing significantly; (2) Stop when parameter estimates in two consecutive iterations no longer differ significantly.

Further on, we present the EM algorithm for solving the optimization problem (2.1).

The available data, called *incomplete data*, are given by the corpus $\mathscr{C}$. The *missing data* are the senses of the ambiguous words, hence they must be modeled by some random variables

$$h_{ik} = \begin{cases} 1, & \text{context } c_i \text{ generates sense } s_k \\ 0, & \text{otherwise} \end{cases}, \quad i = 1, ..., I; \; k = 1, ..., K$$

The *complete data* consist of incomplete and missing data, and the corresponding likelihood of the corpus $\mathscr{C}$ becomes

$$P_{\text{complete}}(\mathscr{C}) = \prod_{i=1}^{I} \prod_{k=1}^{K} \left(\alpha_k \prod_{j=1}^{J} \left(\theta_{kj}\right)^{|v_j \; in \; c_i|} \right)^{h_{ik}}$$

Hence, the log-likelihood for complete data is

$$\log P_{\text{complete}}(\mathscr{C}) = \sum_{i=1}^{I} \sum_{k=1}^{K} h_{ik} \left(\log \alpha_k + \sum_{j=1}^{J} |v_j \; in \; c_i| \cdot \log \theta_{kj} \right)$$

Each M-step of the algorithm solves the maximization problem

$$\max \left(\sum_{i=1}^{I} \sum_{k=1}^{K} h_{ik} \left(\log \alpha_k + \sum_{j=1}^{J} |v_j \; in \; c_i| \cdot \log \theta_{kj} \right) \right) \qquad (2.2)$$

with the constraints

$$\sum_{k=1}^{K} \alpha_k = 1 \qquad \text{for all } k = 1, ..., K$$

$$\sum_{j=1}^{J} \theta_{kj} = 1$$

For simplicity, we denote the vector of parameters by

$$\psi = (\alpha_1, ..., \alpha_K, \theta_{11}, ..., \theta_{KJ})$$

and notice that the number of independent components (parameters) is $(K - 1) + (KJ - K) = KJ - 1$.

The EM algorithm starts with a random initialization of the parameters, denoted by

$$\psi^{(0)} = \left(\alpha_1^{(0)}, ..., \alpha_K^{(0)}, \theta_{11}^{(0)}, ..., \theta_{KJ}^{(0)} \right)$$

The *iteration* $(r + 1)$ consists in the following two steps:

The E-step computes the missing data, based on the model parameters estimated at iteration r, as follows:

$$h_{ik}^{(r)} = P_{\psi^{(r)}} (h_{ik} = 1 \mid \mathscr{C}),$$

$$h_{ik}^{(r)} = \frac{\alpha_k^{(r)} \cdot \prod_{j=1}^{J} \left(\theta_{kj}^{(r)} \right)^{|v_j \text{ in } c_i|}}{\sum_{k=1}^{K} \alpha_k^{(r)} \cdot \prod_{j=1}^{J} \left(\theta_{kj}^{(r)} \right)^{|v_j \text{ in } c_i|}}, \quad i = 1, ..., I; \ k = 1, ..., K$$

The M-step solves the maximization problem (2.2) and computes $\alpha_k^{(r+1)}$ and $\theta_{kj}^{(r+1)}$ as follows:

$$\alpha_k^{(r+1)} = \frac{1}{I} \sum_{i=1}^{I} h_{ik}^{(r)}, \quad k = 1, ..., K$$

$$\theta_{kj}^{(r+1)} = \frac{\sum_{i=1}^{I} |v_j \text{ in } c_i| \cdot h_{ik}^{(r)}}{\sum_{j=1}^{J} \sum_{i=1}^{I} |v_j \text{ in } c_i| \cdot h_{ik}^{(r)}}, \quad k = 1, ..., K; j = 1, ..., J$$

The stopping criterion for the algorithm is "Stop when parameter estimates in two consecutive iterations no longer differ significantly". That is, stop when

$$\left\| \psi^{(r+1)} - \psi^{(r)} \right\| < \varepsilon,$$

namely

$$\sum_{k=1}^{K} \left(\alpha_k^{(r+1)} - \alpha_k^{(r)} \right)^2 + \sum_{k=1}^{K} \sum_{j=1}^{J} \left(\theta_{kj}^{(r+1)} - \theta_{kj}^{(r)} \right) < \varepsilon$$

It is well known that the EM iterations $\left(\psi^{(r)} \right)_r$ converge to the Maximum Likelihood Estimate $\widehat{\psi} = \left(\widehat{\alpha}_1, ..., \widehat{\alpha}_K, \widehat{\theta}_{11}, ..., \widehat{\theta}_{KJ} \right)$.

Once the parameters of the model have been estimated, we can disambiguate contexts of w by computing the probability of each of the senses based on features v_j occurring in the context c. Making the Naïve Bayes assumption and using the Bayes decision rule, we can decide s' if

$$s' = \underset{k=1,...,K}{\arg\max} \left(\log \widehat{\alpha}_k + \sum_{j=1}^{J} \left| v_j \ in \ c \right| \cdot \log \widehat{\theta}_{kj} \right)$$

Our choice of recommending usage of the EM algorithm for parameter estimation in the case of unsupervised WSD with an underlying Naïve Bayes model is based on the fact that this algorithm has proven itself to be not only a successful iterative method, but also one which fits well to models with missing data. However, our choice is based on previously existing discussions and reported disambiguation results as well. The EM algorithm has equally been used for parameter estimation (together with Gibbs sampling), relatively to an underlying Naïve Bayes model, in (Pedersen and Bruce 1998), to the results of which the accuracies obtained by other disambiguation methods (see Chaps. 3 and 5) have constantly been compared. These are disambiguation accuracies resulted when feeding knowledge of completely different natures to the Naïve Bayes model, as a result of using various different ways of performing feature selection (see Chaps. 3–5). The EM algorithm has equally been used with a Naïve Bayes model in (Gale et al. 1995), in order to distinguish city names from people's names. An accuracy percentage in the mid-nineties, with respect to *Dixon*, a name found to be quite ambiguous, was reported.

References

Dempster, A., Laird, N., Rubin, D.: Maximum likelihood from incomplete data via the EM algorithm. J. Roy. Stat. Soc. B **39**(1), 1–38 (1977)

Eberhardt, F., Danks, D.: Confirmation in the cognitive sciences: the problematic case of Bayesian models. Mind. Mach. **21**, 389–410 (2011)

Gale, W., Church, K., Yarowsky, D.: A method for disambiguating word senses in a large corpus. Comput. Humanit. **26**(5–6), 415–439 (1992)

Gale, W.A., Church, K.W., Yarowsky, D.: Discrimination decisions for 100,000-dimensional space. Ann. Oper. Res. **55**(2), 323–344 (1995)

Manning, C., Schütze, H.: Foundations of Statistical Natural Language Processing. The MIT Press, Cambridge (1999)

Pedersen, T., Bruce, R.: Knowledge lean word-sense disambiguation. In: Proceedings of the 15th National Conference on Artificial Intelligence, pp. 800–805. Madison, Wisconsin (1998)

Chapter 3
Semantic WordNet-Based Feature Selection

Abstract The feature selection method we are presenting in this chapter makes use of the semantic network WordNet as knowledge source for feature selection. The method makes ample use of the WordNet semantic relations which are typical of each part of speech, thus placing the disambiguation process at the border between unsupervised and knowledge-based techniques. Test results corresponding to the main parts of speech (nouns, adjectives, verbs) will be compared to previously existing disambiguation results, obtained when performing a completely different type of feature selection. Our main conclusion will be that the Naïve Bayes model reacts well in the presence of semantic knowledge provided by WN-based feature selection when acting as clustering technique for unsupervised WSD.

Keywords Bayesian classification · Word sense disambiguation · Unsupervised disambiguation · Knowledge-based disambiguation · WordNet

3.1 Introduction

The feature selection method we shall be describing here makes use of the semantic network WordNet for creating a disambiguation vocabulary that contains a restricted number of words (features) for unsupervised WSD with an underlying Naïve Bayes model. The number of parameters which are to be estimated by the EM algorithm[1] is therefore reduced by performing knowledge-based feature selection. The novelty of the presented method consists in using the semantic network WordNet as knowledge source for feature selection. The method makes ample use of the WordNet semantic relations which are typical of each part of speech (see Sect. 3.3) and therefore places the disambiguation process at the border between unsupervised and knowledge-based techniques. Test results corresponding to all major parts of speech (nouns, adjectives, verbs) have been performed (Hristea et al. 2008; Hristea 2009; Hristea and Popescu 2009) and have shown that feature selection using a knowledge source

---

[1] See the mathematical model presented in Chap. 2.

F. T. Hristea, *The Naïve Bayes Model for Unsupervised Word Sense Disambiguation*,   17
SpringerBriefs in Statistics, DOI: 10.1007/978-3-642-33693-5_3, © The Author(s) 2013

of type WordNet is more effective in disambiguation than local-type features (like part-of-speech tags) are.

3.2 WordNet

WordNet (Miller 1990, 1995; Miller et al. 1990) is a large electronic and interactive lexical database for English. It has been developed during the last 20 years[2] at Princeton University by a group headed by Professor George Miller, a psycholinguist who was inspired by experiments in Artificial Intelligence that tried to understand human semantic memory [e.g., (Collins and Quillian 1969)[3]]. The novelty of Miller's approach was the attempt to represent the entire bulk of the lexicalized concepts of a language in a network-like structure based on hierarchical relations.

As its authors note, WordNet (WN)[4] is a lexical knowledge base which was created as a machine-readable dictionary based on psycholinguistic principles. It is a lexical database that currently contains (ver. 3.0) approximately 155,287 English nouns, verbs, adjectives and adverbs organized by semantic relations into over 117,000 meanings, where a meaning is represented by a set of synonyms (a *synset*) that can be used (in an appropriate context) to express that meaning. These numbers are approximate since WN continues to grow. The building block of WN is the synset. A synset lexically expresses a *concept*. A word's membership in multiple synsets reflects that word's polysemy. Different relations link the WN synsets. An entry in WN consists of a synset, a definitional gloss, and (sometimes) one or more phrases illustrating usage. The major relations used to organize words and entries are synonymy and antonymy, hyponymy, troponymy and hypernymy, meronymy and holonymy.

WN was primarily viewed as a lexical database. However, due to its structure, it can be equally considered a semantic network and a knowledge base. It has been recognized as a valuable resource in the human language technology and knowledge processing communities. In WSD, WN represents the most popular sense inventory, with its synsets being used as labels for sense disambiguation.

WSD can be performed at many levels of granularity. The various existing sense inventories have different such levels of granularity. WN is very fine-grained, while other possible sense inventories, such as thesauri and dictionaries, have much lower granularity. The level of granularity offered by the sense inventory has great influence

---

[2] In 1986 George Miller has the initiative of creating WordNet and designs its structure, which was meant to serve testing current theories concerning human semantic memory. Verbs are added to the network the following year (1987) and its first version (1.0) is released in 1991. Already in 2006 approximately 8000 download operations were registered on a daily basis and similar, more or less developed, semantic networks of type WordNet existed for some 40 other languages.

[3] The Collins and Quillian model proposed a hierarchical structure of concepts, where more specific concepts inherit information from their superordinate, more general concepts. That is why only knowledge particular to more specific concepts needs to be stored with such concepts.

[4] For a comprehensive description of WN see also Fellbaum (1998).

over WSD, making the problem more or less difficult, and is therefore taken into account in the evaluation of WSD systems. As already mentioned, WN has become the most popular sense inventory nowadays.

As the American WN continues to grow, new features are added to it. Version 2.1, for instance, is the first to incorporate the distinctions between classes and instances reported in (Miller and Hristea 2006) which lead to a semi-ontology of WN nouns. And which facilitate the disambiguation of proper names.

According to specific semantic relations, in WN, noun and verb synsets are organized as hierarchies, while adjective and adverb synsets are part of a completely different structure—the *cluster*.

Let us finally note that WN represents *words* and *concepts* as an interrelated system which, according to Miller (1998), is consistent with evidence of the way speakers organize their mental lexicons. And which incorporates knowledge into the lexicon, bringing it closer to the mental one, that contains both word and world knowledge (Kay 1989). This should be of the essence for artificial intelligence applications such as WSD and is in contrast to linguistic theories that attempt to model human grammar, or linguistic competence. Unlike such linguistic theories (as the one used in Chap. 4), the structure of WordNet is motivated by theories of human knowledge organization (Fellbaum 1998, p. 2).

The Naïve Bayes model has been shown (Hristea et al. 2008; Hristea 2009; Hristea and Popescu 2009) to react well, from the point of view of unsupervised WSD, in the presence of knowledge such as that offered by WordNet.

3.3 Making Use of WordNet for Feature Selection

The approach to WSD, more precisely to word sense discrimination, of Hristea et al. (2008), which we are describing here, relies on a set of features formed by the actual words occurring near the target word (within the context window) and tries to reduce the size of this feature set by performing knowledge-based feature selection. The semantic network WordNet has been used as unique knowledge source for feature selection. While the classical approach forms the vocabulary on which the disambiguation process relies dynamically, using all the content words which occur in the contexts, the present approach forms the same vocabulary based entirely on WordNet. The WN semantic network will provide the words considered relevant for the set of senses taken into consideration corresponding to the target word.

First of all, words occurring in the same WN synsets as the target word (WN synonyms) have been chosen (Hristea et al. 2008), corresponding to all senses of the target. Additionally, the same authors consider the words occurring in synsets related (through explicit relations provided in WN) to those containing the target word as part of the vocabulary used for disambiguation. Synsets and relations were restricted to those associated with the part of speech of the target word. The content words of the glosses of all types of synsets participating in the disambiguation process, using the example string associated with the synset gloss as well, were equally taken into consideration (Hristea et al. 2008). The latter choice was made since previous studies

(Banerjee and Pedersen 2003), performed for knowledge-based disambiguation, have come to the conclusion that the *example relation*—which simply returns the example string associated with the input synset—seems to provide useful information in the case of all parts of speech. A conclusion which is not surprising, as the examples contain words related syntagmatically to the target.

With respect to nouns, which represent the best developed portion of WordNet, previous studies (Banerjee and Pedersen 2003), performed for knowledge-based disambiguation, come to the conclusion that *hyponym* and *meronym* synsets are the most informative. However, in (Hristea et al. 2008) *hypernyms* and *holonyms* are equally taken into consideration. Tables 3.5 and 3.7 of Sect. 3.4.2 show the obtained disambiguation results when using various combinations of the mentioned types of WN synsets in the formation of the "disambiguation vocabulary".

Corresponding to adjectives, the discussed disambiguation method has taken into account (Hristea et al. 2008; Hristea and Popescu 2009) the *similarity* relation, which is typical of adjectives (and, in fact, only holds for adjective synsets contained in adjective clusters[5]). The *also-see* relation and the *attribute* relation have also been taken into account since these relations are considered most informative and have been found (Banerjee and Pedersen 2003) to rank highest among the useful relations for adjectives. The *pertaining-to* relation has also been considered, whenever possible. Finally, the *antonymy* relation has represented a source of "negative information" that has proven itself useful in the disambiguation process. This is in accordance with previous findings of studies performed for knowledge-based disambiguation (Banerjee and Pedersen 2002) that consider the antonymy relation a source of negative information allowing a disambiguation algorithm "to identify the sense of a word based on the absence of its antonymous sense in the window of context". Tables 3.8 and 3.9 of Sect. 3.4.2 show the obtained disambiguation results (Hristea et al. 2008; Hristea and Popescu 2009) when using a disambiguation vocabulary in the formation of which all mentioned types of synsets have taken part. This is the vocabulary which has provided the best disambiguation results in the case of adjectives *common* and *public* (see Sect. 3.4.2.2). Disambiguation results were computed with and without antonym synsets participating in the disambiguation process.

In the case of verbs, it has been suggested (Hristea et al. 2008; Hristea 2009) to additionally use, whenever possible, WN synsets indicated by the *entailment relation*[6] and by the *causal relation*,[7] which are typical of this part of speech. Table 3.10

---

[5] WordNet divides adjectives into two major classes: descriptive and relational. Descriptive adjectives are organized into clusters on the basis of binary opposition (antonymy) and similarity of meaning (Fellbaum 1998). Descriptive adjectives that do not have direct antonyms are said to have indirect antonyms by virtue of their semantic similarity to adjectives that do have direct antonyms. Relational adjectives are assumed to be stylistic variants of modifying nouns and are cross-referenced to the noun files (see the relation "relating-or-pertaining-to"). The function such adjectives play is usually that of classifying their head nouns (Fellbaum 1998).

[6] The entailment relation between verbs resembles meronymy between nouns, but meronymy is better suited to nouns than to verbs (Fellbaum 1998).

[7] The causal relation (Fellbaum 1998) picks out two verb concepts, one causative (like *give*), the other what might be called the "resultative" (like *have*).

of Sect. 3.4.2 shows the obtained disambiguation results (Hristea et al. 2008) in the case of verb *help*.

As a result of using only those words indicated as being relevant by WordNet, a much smaller vocabulary was obtained, and therefore a much smaller number of features have taken part in the disambiguation process. In the case of this method each word (feature) contributes to the final score being assigned to a sense with a weight given by $P(v_j \mid s_k)$.[8] This weight (probability) is not a priori established, but is learned by means of the EM algorithm.

3.4 Empirical Evaluation

Tests concerning the described disambiguation method have initially concentrated on adjectives (Hristea and Popescu 2009). An experiment concerning verbs has also been performed (Hristea 2009). The method was extended to nouns and surveyed in (Hristea et al. 2008), where conclusions regarding all these parts of speech were presented.

In the case of nouns, the part of speech for which the best disambiguation results had been recorded by the literature, the goal of the performed experiment (Hristea et al. 2008) was to compare results obtained by means of the new disambiguation method with those obtained by a classical unsupervised algorithm (one having an underlying Naïve Bayes model, which does not perform feature selection and which is trained with the EM algorithm). The obtained disambiguation results (Hristea et al. 2008) were equally compared to those of Pedersen and Bruce (1998), where an algorithm of the same type (unsupervised with an underlying Naïve Bayes model) is placed under survey. However, the algorithm studied by Pedersen and Bruce relies on a restricted set of local features, that include co-occurrence and part of speech information near the target word (as commented in Chap. 1). It therefore performs feature selection, although in a completely different manner than that proposed by the described method. With the necessity of performing feature selection of some type being obvious, disambiguation results concerning adjectives and verbs were compared with those of Pedersen and Bruce (1998) only. In the case of all parts of speech test results have shown (Hristea et al. 2008) that feature selection using a knowledge source of type WordNet is more effective in sense disambiguation than local-type features are.

3.4.1 Design of the Experiments

In what follows, we are describing the experiments designed in (Hristea et al. 2008), experiments which have led to the conclusion that semantic WN-based features are more effective in sense discrimination than local type ones.

---

[8] See the mathematical model presented in Chap. 2.

Table 3.1 Distribution
of senses of *line*

Sense	Count	
Product	2,218	(53,47%)
Written or spoken text	405	(9,76%)
Telephone connection	429	(10,34%)
Formation of people or things; queue	349	(8,41%)
An artificial division; boundary	376	(9,06%)
A thin, flexible object; cord	371	(8,94%)
Total count	4,148	

3.4.1.1 Noun Experiment

In the case of nouns, the *line* corpus (Leacock et al. 1993) has been used (Hristea et al. 2008) as test data. This corpus contains around 4,000 examples of the word *line* (noun) sense-tagged with subsets of their WordNet 1.5 senses. Examples are drawn from the WSJ corpus, the American Printing House for the Blind, and the San Jose Mercury. The *line* data set was chosen (Hristea et al. 2008) for tests concerning nouns since it seems to have raised the greatest problems in the case of the Pedersen and Bruce (1998) approach to WSD, to which the results of the new method were compared. Pedersen and Bruce obtain the most modest disambiguation results in the case of the noun *line* (when testing for 5 different nouns).

The *line* data was created by Leacock et al. (1993) by tagging every occurrence of *line* in the selected corpus with one of 6 possible WordNet senses. These senses and their frequency distribution are shown in Table 3.1.

In order for the experiments to be conducted, the *line* corpus was preprocessed in the usual required way for WSD: the stop words were eliminated, and Porter stemmer was applied to the remaining words.

Two types of tests were performed in the case of the classical unsupervised algorithm. A first variant of testing involved a context window of size 5, which is a common size for WSD tests of this type. The second testing variant used a context window of size 25. This dimension was chosen in order for the two methods (the classical one and the newly proposed one) to be compared under the same conditions.

The newly introduced method generated a series of experiments that vary according to the specific sources used in establishing the so-called disambiguation vocabulary.

The overall source for creating the vocabulary was WordNet 3.0, which lists 30 different senses corresponding to the noun *line*. The *line* corpus is sense-tagged with subsets of WordNet 1.5 senses, namely with those senses listed in Table 3.1. Therefore a sense mapping of the initial (corpus) senses to those of the WN 3.0 database was necessary.

The sense "product" occurring in the *line* corpus has been mapped to the WN 3.0 synset having the *synset_id* 103671668 and containing the nouns {line, product line, line of products, line of merchandise, business line, line of business}. The sense "written or spoken text" occurring in the corpus corresponds to 3 WN 3.0 synsets, namely synset {note, short letter, line, billet} having the *id* 106626286, synset {line} having the *id* 107012534, and synset {line} having the *id* 107012979, respectively. The sense "telephone connection" occurring in the corpus corresponds to the WN 3.0 synset {telephone line, phone line, telephone circuit, subscriber line, line} having the *id* 104402057. The sense "formation of people or things; queue" occurring in the corpus corresponds to 2 WN 3.0 synsets, namely synset {line} having the *id* 108430203 and synset {line} having the *id* 108430568, respectively. The sense "an artificial division; boundary" occurring in the corpus corresponds to the WN 3.0 synset {line, dividing line, demarcation, contrast} having the *id* 105748786. Finally, the sense "a thin, flexible object; cord" occurring in the corpus corresponds to the WN 3.0 synset {line} having the *id* 103670849.

Let us once again note that this disambiguation method is an unsupervised one and therefore does not require sense labels (but only the number of senses, as detailed in Sect. 3.4.2). Performing the presented sense mapping was necessary solely for establishing the restricted disambiguation vocabulary (relevant words).

Once the subset of senses taking part in the experiments had been established, the relevant information for building the vocabulary had to be specified.

The first performed experiment involving the disambiguation of the noun *line* establishes (Hristea et al. 2008) as relevant words forming the vocabulary all nouns of the 9 WN 3.0 synsets containing *line* which have been chosen as a result of sense mapping. Additionally, all content words occurring in the glosses of these synsets have been added to this vocabulary.[9] Within the following experiments information provided by the synsets related (through explicit relations existing in WN) to those containing the target word *line* has been successively added. Thus, the second performed experiment uses, along with all words occurring in the first one, the words existing in the hyponym and meronym synsets of the 9 synsets containing the target.[10] The third experiment uses all words occurring in the second one, to which all content words of all the hyponym and meronym synset glosses are added.[11] Within the next experiment the initially used vocabulary (first experiment) has been enriched by adding all words coming from all hyponym, hypernym, meronym and holonym synsets of the 9 synsets containing the target.[12] Finally, the last experiment uses all words involved in the previously described one, to which the content words occurring in the glosses of all synsets required by the previous experiment are added.[13]

3.4.1.2 Adjective Experiment

In the case of adjectives, in (Hristea et al. 2008) the test data is represented by the Bruce et al. (1996) data containing twelve words taken from the ACL/DCI Wall Street

---

[9] Experiment referred to in Tables 3.5 and 3.7 as "Synonyms + Glosses".

[10] Experiment referred to in Tables 3.5 and 3.7 as "+Hyponyms + Meronyms".

[11] Experiment referred to in Tables 3.5 and 3.7 as "+Hyponyms + Glosses + Meronyms + Glosses".

[12] Experiment referred to in Tables 3.5 and 3.7 as "+Hyponyms + Hypernyms + Meronyms + Holonyms".

[13] Experiment referred to in Tables 3.5 and 3.7 as "+Hyponyms + Glosses + Hypernyms + Glosses + Meronyms + Glosses + Holonyms + Glosses".

Table 3.2 Distribution
of senses of *common*

Sense	Count
As in the phrase "common stock"	84 %
Belonging to or shared by 2 or more	8 %
Happening often; usual	8 %
Total count	1,060

Table 3.3 Distribution
of senses of *public*

Sense	Count
Concerning people in general	68 %
Concerning the government and people	19 %
Not secret or private	13 %
Total count	715

Journal corpus and tagged with senses from the Longman Dictionary of Contemporary English. This data set was chosen for tests concerning adjectives since it has equally been used in the case of the Pedersen and Bruce (1998) approach to WSD, to which the results reported in (Hristea et al. 2008) were constantly compared.

Test results have been reported in the case of two adjectives, *common* and *public*, the latter being the one corresponding to which Pedersen and Bruce obtain the most modest disambiguation results. The senses of *common* that have been taken into consideration and their frequency distribution are shown in Table 3.2, while Table 3.3 provides the same type of information corresponding to the adjective *public*. In these tables *total count* represents the number of occurrences in the corpus of each word, with each of the adjectives being limited to the 3 most frequent senses, while *count* gives the percentage of occurrence corresponding to each of these senses. In fact, the choice of performing tests in the case of adjectives *common* and *public* has been influenced (Hristea et al. 2008) by the fact that these adjectives are represented in the mentioned corpus by three different senses, while the other two adjectives for which Pedersen and Bruce perform disambiguation tests, *chief* and *last*, have only two senses (in the same corpus). Since unsupervised disambiguation should be able to produce distinctions even between usage types that are more fine grained than would be found in a dictionary, the choice of testing in the case of those adjectives having the greatest number of senses represented in the corpus becomes a natural one.

In order for the experiments to be conducted, the data set was preprocessed in the usual required way for WSD: the stop words were eliminated, and Porter stemmer was applied to the remaining words.

The overall source for creating the disambiguation vocabulary (Hristea et al. 2008) was again WordNet 3.0, which lists 9 different senses corresponding to the adjective *common* and only 2 different senses corresponding to the adjective *public*. Obviously, a sense mapping of the initial (corpus) senses to those of the WN 3.0 database was again necessary. According to this mapping, 4 WN 3.0 synsets took part in the disambiguation vocabulary corresponding to the adjective *common*, namely the synsets having the IDs 300492677,[14] 302152473,[15] 301673815[16] and 300970610,[17] respectively. Both WN synsets corresponding to the adjective *public* and having the IDs 300493297[18] and 301861205,[19] respectively were part of the same vocabulary when performing disambiguation tests relative to this adjective.

Once the subset of WN senses taking part in the experiments was established, the relevant information for building the disambiguation vocabulary had to be specified once again.

Each of the experiments involving the disambiguation of adjectives *common* and *public* have established (Hristea et al. 2008; Hristea and Popescu 2009) as relevant words forming the vocabulary all words of the WN 3.0 synsets containing the respective adjective which have been chosen as a result of sense mapping. Additionally, all content words occurring in the glosses and the associated example strings of these synsets have been added to this vocabulary. Information provided by the synsets related (through explicit relations existing in WN) to those containing the target word has also been included in the same vocabulary. Thus, the first performed experiment[20] additionally uses all content words occurring in the synsets, their corresponding glosses and example strings, given by the similarity relation, the also-see relation, the attribute relation, the pertaining-to relation, whenever possible, and, finally, the antonymy relation, which has been considered interesting due to the "negative information" it can provide. The second performed experiment[21] eliminates from the disambiguation vocabulary all words brought in precisely by these antonym synsets.

3.4.1.3 Verb Experiment

The newly proposed disambiguation method has been tested (Hristea et al. 2008; Hristea 2009) in the case of verbs as well, since it is well known that the verb represents the part of speech which is the most difficult to disambiguate. Test results were equally compared to those obtained in (Pedersen and Bruce 1998). Corresponding to verbs the Bruce et al. (1996) data was used as test data once again. From this 12-word sense-tagged corpus the verb *help* was selected, out of a total of 4 sense-tagged verbs. This choice was again determined by the fact that *help* is a verb in the case of which disambiguation results are quite modest when using the Pedersen-and-Bruce-type local features. The distribution of senses corresponding to *help* that has been used

---

[14] This is synset {common} having the gloss 'belonging to or participated in by a community as a whole; public'.

[15] This is synset {common, mutual} having the gloss 'common to or shared by two or more parties'.

[16] This is synset {common} having the gloss 'to be expected; standard'.

[17] This is synset {common, usual} having the gloss 'commonly encountered'.

[18] This is synset {public} having the gloss 'affecting the people or community as a whole'.

[19] This is synset {public} having the gloss 'not private; open to or concerning the people as a whole'.

[20] Referred to in Tables 3.8 and 3.9 as "all".

[21] Referred to in Tables 3.8 and 3.9 as "all-antonyms".

Table 3.4 Distribution
of senses of *help*

Sense	Count
To enhance-inanimate object	78 %
To assist-human object	22 %
Total count	1,267

in the performed experiments, as well as in (Pedersen and Bruce 1998), is shown in
Table 3.4.

The discussed disambiguation method has again generated a series of experiments
that vary according to the specific sources used in establishing the disambiguation
vocabulary. The overall source for creating this vocabulary was again WordNet 3.0,
which lists 8 different senses corresponding to the verb *help*. In the case of this verb
the disambiguation vocabulary was formed by taking into account all verbs of the
6 WN 3.0 synsets[22] containing *help* which have been chosen as a result of sense
mapping, all content words occurring in the glosses and the associated example
strings of these synsets, as well as all content words belonging to all WN-related
synsets, their glosses and their corresponding example strings. This vocabulary is
regarded (Hristea et al. 2008; Hristea 2009) as an extended one, thus created in order
to ensure greater coverage of the corpus instances for participation in the learning
process, a requirement which is always more difficult to meet in the case of verbs.

3.4.2 Test Results

In all mentioned studies (Pedersen and Bruce 1998; Hristea et al. 2008; Hristea
2009; Hristea and Popescu 2009) concerning unsupervised WSD with an underlying
Naïve Bayes model, performance is evaluated in terms of accuracy. In the case of
unsupervised disambiguation, however, defining accuracy is not as straightforward
as in the supervised case. The objective is to divide the I given instances of the
ambiguous word into a specified number K of sense groups, which are in no way
connected to the sense tags existing in the corpus. In the performed experiments, sense
tags are used only in the evaluation of the sense groups found by the unsupervised
learning procedure. These sense groups must be mapped to sense tags in order to

---

[22] These are the following:

- synset {help, aid} having the ID 200082081 and the gloss 'improve the condition of';
- synset {help} having the ID 200206998 and the gloss 'improve; change for the better';
- synset {serve, help} having the ID 201181295 and the gloss 'help to some food; help with food or drink';
- synset {avail, help} having the ID 201193569 and the gloss 'take or use';
- synset {help, assist, aid} having the ID 202547586 and the gloss 'give help or assistance; be of service';
- synset {help} having the ID 202555434 and the gloss 'contribute to the furtherance of'.

evaluate system performance. The mapping that results in the highest classification accuracy[23] has been used.

Test results are presented in Tables 3.5, 3.7, 3.8, 3.9 and 3.10.[24] Each result represents the average accuracy and standard deviation obtained by the learning procedure over 20 random trials while using a context window of size 25[25] and a threshold ε having the value 10^{-9}. Tables 3.5 and 3.7 (corresponding to nouns) also present, for enabling comparison, the results obtained by the classical algorithm, when using a context window of size 5 and 25, respectively. These results equally represent the average accuracy and standard deviation over 20 trials of the EM algorithm with a threshold ε having the value 10^{-9}.

Apart from accuracy, the following type of information is also included in Tables 3.5, 3.7, 3.8, 3.9 and 3.10: number of features resulting in each experiment and percentage of instances having only null features (i.e. containing no relevant information).

As previously mentioned, within the present approach to disambiguation, the value of a feature is given by the number of occurrences of the corresponding word in the given context window. Since the process of feature selection is based on the restriction of the disambiguation vocabulary, it is possible for certain instances not to contain (in their context window) any of the relevant words forming this vocabulary. Such instances will have null values corresponding to all features. The smaller the number of features used for disambiguation, the more frequently this takes place. These instances do not contribute to the learning process. However, they have been taken into account in the evaluation stage of the presented experiments. Corresponding to these instances, the algorithm assigns the sense s_k for which the value of $P(s_k)$ (estimated by the EM algorithm)[26] is maximal.

3.4.2.1 Test Results Concerning Nouns

In the case of nouns, as can be seen in Table 3.5, the obtained disambiguation results (Hristea et al. 2008) when using an underlying Naïve Bayes model and applying the EM algorithm to the classical set of features, formed with the actual words occurring

---

[23] In order to conduct their experiments the mentioned authors have chosen a number of sense groups equal to the number of sense tags existing in the corpus. Therefore a number of $K!$ possible mappings (with K denoting the number of senses of the target word) should be taken into account. For a fixed mapping, its accuracy is given by the number of correct labellings (identical to the corresponding corpus sense tags) divided by the total number of instances. From the $K!$ possible mappings, the one with maximum accuracy has been chosen.

[24] Reprinted here from (Hristea et al. 2008).

[25] The choice of this context window size is based on the suggestion of Lesk (1986) that the quantity of data available to the algorithm is one of the biggest factors to influence the quality of disambiguation. In this case, a larger context window allows the occurrence of a greater number of WN relevant words (with respect to the target), which are the only ones to participate in the creation of the disambiguation vocabulary.

[26] See the mathematical model presented in Chap. 2.

Table 3.5 Experimental results for 6 senses of *line*

Method	No. of features	Percentage of instances having only null features	Accuracy
Classic-5	4700	0.0	0.274 ± 0.02
Classic-25	9932	0.0	0.255 ± 0.02
Synonyms + Glosses	73	45.7	0.473 ± 0.02
+ hyponyms + meronyms	138	38.3	0.478 ± 0.01
+ hyponyms + glosses + meronyms + glosses	305	11.7	0.454 ± 0.04
+ hyponyms + hypernyms + meronyms + holonyms	152	35.9	0.465 ± 0.02
+ hyponyms + glosses + hypernyms + glosses + meronyms + glosses + holonyms + glosses	358	8.5	0.448 ± 0.05

in the context window, are extremely modest. A possible cause of this failure is the great number of features used by the learning algorithm. This is also suggested by the fact that, when enlarging the context window from size 5 to size 25, the number of features increases from 4,700 to 9,932, which leads to a decrease in accuracy from 0.274 to 0.255. Let us note that accuracy in the same range (25–30 %) is reported in (Pedersen and Bruce 1998) when tests corresponding to all 6 senses of the *line* corpus are performed.

The first conclusion that results presented in Table 3.5 immediately lead to is that, whenever performing feature selection, accuracy increases substantially.

The best disambiguation result (0.478) was obtained (Hristea et al. 2008) in the case when the disambiguation vocabulary was formed with all WN synonyms occurring in all synsets that contain the target word, content words of the glosses corresponding to these synsets, as well as nouns coming from all their hyponym and meronym synsets. This is in accordance with previous findings of studies performed for knowledge-based disambiguation (Banerjee and Pedersen 2003) concluding that, in the case of nouns, hyponym and meronym synsets of those containing the target word are the most informative. The obvious conclusion is that making use of a knowledge base of type WordNet (in this case, for feature selection) substantially improves disambiguation results.

The same set of experiments, performed under the same conditions, has been conducted (Hristea et al. 2008) in the case of only 3 senses of *line*. The reason for performing this reduction from 6 to 3 senses was to verify to what extent the existence of a majority sense in the distribution of senses for *line*[27] influences the

---

[27] Sense "product" occurs in 53,47 % of the *line* corpus instances.

Sense	Count
Telephone connection	429 (37,33%)
Formation of people or things; queue	349 (30,37%)
A thin, flexible object; cord	371 (32,28%)
Total count	1,149

Table 3.6 Distribution of the 3 chosen senses of *line*

performances of the presented disambiguation method. The 3 chosen senses are listed in Table 3.6.[28]

A more uniform distribution of the *line* senses has thus been obtained. Additionally, the 3 senses that have been selected (Hristea et al. 2008) coincide with the ones used in (Pedersen and Bruce 1998) for the presented disambiguation experiment, a choice which has allowed a straightforward comparison between the corresponding results.

Test results for this case (3 senses of *line*) are presented in Table 3.7. As in the previous case (6 senses of *line*) the results obtained by the classical algorithm (without feature selection) are very modest, while performing feature selection consistently improves the results corresponding to each experiment. The maximum obtained accuracy (0.591) represents a consistent improvement over the maximum obtained in the previous case (0.478). This maximum accuracy was obtained corresponding to a disambiguation vocabulary formed with all words occurring in all synsets containing the target word, their hyponym, hypernym, meronym, and holonym synsets, to which all content words of all corresponding glosses were added. The explanation for obtaining the best result when using a larger number of explicit relations provided in WordNet could reside in the fact that, corresponding to the three chosen senses of *line*, no meronym synsets exist. This considerably reduces the disambiguation vocabulary that resulted in the best accuracy when all 6 senses of *line* were disambiguated.

The presented disambiguation method and corresponding results have been compared (Hristea et al. 2008) primarily to those of Pedersen and Bruce (1998) since both methods rely on an underlying Naïve Bayes model, use the EM algorithm for estimating model parameters[29] in unsupervised WSD and perform feature selection. The main difference between the two approaches consists in the way feature selection is performed. While Pedersen and Bruce, as mentioned before, use local features that include co-occurrence and part of speech information near the target word, the present approach relies on WordNet and its rich set of semantic relations for performing feature selection. This places the disambiguation process at the border between unsupervised and knowledge-based techniques, but improves disambiguation accuracy consistently. Thus, the way in which this method performs feature selection brings the same disambiguation accuracy when testing for all 6 senses of *line* as that obtained in (Pedersen and Bruce 1998) in the case of only 3 chosen senses of this target word (47%). The mentioned authors report that "accuracy degrades

[28] Reprinted here from (Hristea et al. 2008).
[29] Pedersen and Bruce (1998) also make use of Gibbs sampling for parameter estimation, without results improving significantly.

Table 3.7 Experimental results for 3 senses of *line*

Method	No. of features	Percentage of instances having only null features	Accuracy
Classic-5	1907	0.0	0.280 ± 0.02
Classic-25	4806	0.0	0.248 ± 0.02
Synonyms + Glosses	30	49.7	0.487 ± 0.03
+ hyponyms + meronyms	74	41.7	0.513 ± 0.03
+ hyponyms + glosses + meronyms + glosses	203	17.9	0.570 ± 0.08
+ hyponyms + hypernyms + meronyms + holonyms	82	38.5	0.498 ± 0.03
+ hyponyms + glosses + hypernyms + glosses + meronyms + glosses + holonyms + glosses	229	15.1	0.591 ± 0.06

Table 3.8 Experimental results for 3 senses of *common*

Method	No. of features	Percentage of instances having only null features	Accuracy
All	83	19.2	0.775 ± 0.02
All—antonyms	74	20.0	0.766 ± 0.04

considerably, to approximately 25–30 %, depending on the feature set" when testing for all 6 senses taken into consideration in the *line* corpus. When disambiguating only the same three chosen senses of *line*, the accuracy of the discussed method is significantly higher (59 %), being obtained with significant corpus coverage (the percentage of instances having only null features is 15.1). This clearly shows that feature selection using a knowledge source of type WordNet can be more effective in disambiguation than local-type features (like part-of-speech tags).

3.4.2.2 Test Results Concerning Adjectives

The presented method has been tested (Hristea et al. 2008; Hristea and Popescu 2009) with respect to adjectives *common* and *public*, the latter being the one in the case of which Pedersen and Bruce (1998) obtain the most modest disambiguation results. Test results are presented in Table 3.8 (corresponding to adjective *common*) and in Table 3.9 (corresponding to adjective *public*).

It can be noticed that the way in which this method performs feature selection brings a disambiguation accuracy of 0.775±0.02 in the case of the adjective *common*, while the highest accuracy obtained in (Pedersen and Bruce 1998), corresponding to the same adjective and when estimating model parameters with the EM algorithm as

Table 3.9 Experimental results for 3 senses of *public*

Method	No. of features	Percentage of instances having only null features	Accuracy
All	74	43.3	0.559 ± 0.03
All—antonyms	71	44.4	0.550 ± 0.03

well, is of 0.543 ± 0.09. When leaving out antonym synsets the accuracy obtained by the discussed method decreases to 0.766 ± 0.04, which again represents a value significantly higher than the corresponding one of Pedersen and Bruce (1998). In the case of adjective *public* this method attains an accuracy of 0.559 ± 0.03, which decreases to 0.550 ± 0.03 when leaving out antonym synsets, with both values being higher than the corresponding one obtained in (Pedersen and Bruce 1998): 0.507 ± 0.03. These results clearly show that feature selection using a knowledge source of type WordNet can be more effective in disambiguation than local-type features (like part-of-speech tags).

When analyzing the results presented in Tables 3.8 and 3.9 one must also notice that accuracy decreases each time the information provided by the antonym synsets is left out of the disambiguation vocabulary. Although there is an obviously restricted number of antonym synsets (see the number of features in the tables) the type of negative information they provide seems to be beneficial to the disambiguation process.

Finally, the fact that, although adjective *public* has only two senses in WN 3.0, discrimination among three different senses was possible, reinforces the idea that unsupervised WSD in general, and that based on an underlying Naïve Bayes model in particular, is able to make distinctions between very fine grained usage types, even more fine grained than those present in a knowledge source of type WordNet.

3.4.2.3 Test Results Concerning Verbs

Usage of the discussed disambiguation method has been exemplified and examined (Hristea et al. 2008; Hristea 2009) in the case of verb *help*, corresponding to which Pedersen and Bruce obtain the most modest results (when estimating model parameters by means of the EM algorithm), out of 4 studied verbs. In the case of *help*, the best disambiguation accuracy attained in (Pedersen and Bruce 1998) is 0.602 ± 0.03. Since, in this case, all WN synonyms corresponding to all chosen synsets will be verbs, which are unlikely to have multiple occurrences in the context window of the target word, and in order to ensure greater coverage of the corpus instances for participation in the learning process, an "extended disambiguation vocabulary" has been taken into account (Hristea et al. 2008; Hristea 2009) as mentioned in Sect. 3.4.1.3. This vocabulary was created by using all verbs of the 6 WN 3.0 synsets containing *help* that have resulted after performing sense mapping, all content words occurring in the glosses and the associated example strings of these synsets, as well as all content words belonging to all WN-related synsets, their glosses and their

Table 3.10 Experimental
results for 2 senses of *help*

Method	No. of features	Percentage of instances having only null features	Accuracy
All	130	41.8	0.671 ± 0.04

corresponding example strings. The presented disambiguation method was asked to perform discrimination among the two senses of *help* chosen in (Pedersen and Bruce 1998) and presented in Table 3.4. As shown in Table 3.10, the obtained accuracy was 0.671 ± 0.04 with a number of 130 resulting features and with over 50 % of the instances contributing to the learning process. This represents an improvement of the result obtained in (Pedersen and Bruce 1998), although the verb is the most difficult to disambiguate part of speech.

Additionally, it is our belief that disambiguation results will improve corresponding to those verbs for which related synsets via the entailment and the causal relations, which are typical of verbs, exist. This was not the case of *help*, corresponding to which only hyponym and hypernym synsets were found. However, *help* was chosen for the performed experiments since it enables comparison with results found in (Pedersen and Bruce 1998). In the case of verbs as well, feature selection using a knowledge source of type WordNet has once again proven to be more effective in disambiguation than local-type features are.

One of the main problems which persists, when using the presented disambiguation method in the case of verbs, is that of low corpus coverage. In an attempt to deal with this problem and for the purpose of the present discussion, we have performed an experiment which extends the disambiguation vocabulary even more. Thus, we have enriched the disambiguation vocabulary used in the experiment presented in Table 3.10 by adding to it the first 1 % of the *most frequent words* which occur in the context windows of the target. In the case of *help* this gives us 50 new words which have been added to the 130 features indicated by WordNet. Two of these words coincide with existing ones, which leaves us with 178 features. In this case, the accuracy of the described method decreases from 0.671 ± 0.04 to 0.602 ± 0.04. The latter result is extremely close to the corresponding (Pedersen and Bruce 1998) accuracy (0.602 ± 0.03). This decrease in accuracy (despite the fact that the most frequently occurring words have been used as features) reinforces the idea that it is the power of WordNet's semantic relations that helps the Naïve Bayes model when acting as clustering technique for unsupervised WSD.

3.5 Conclusions

We have presented a relatively new word sense disambiguation method (Hristea et al. 2008; Hristea 2009; Hristea and Popescu 2009) that lies at the border between unsupervised and knowledge-based techniques. The method performs unsupervised word sense disambiguation based on an underlying Naïve Bayes model, while using WordNet as knowledge source for feature selection. The performance of the method has been compared (Hristea et al. 2008; Hristea 2009; Hristea and Popescu 2009) to that of a previous approach that relies on completely different feature sets (Pedersen and Bruce 1998). Test results for all involved parts of speech have shown that feature selection using a knowledge source of type WordNet is more effective in disambiguation than local-type features (like part-of-speech tags) are. The presentation of the method has reinforced the benefits of combining the unsupervised approach to the WSD problem with a knowledge source of type WordNet. Our main conclusion is that the Naïve Bayes model reacts well in the presence of semantic knowledge provided by WN-based feature selection when acting as clustering technique for unsupervised WSD.

References

Banerjee, S., Pedersen, T.: An adapted Lesk algorithm for word sense disambiguation using Word-Net. In: Proceedings of the Third International Conference on Intelligent Text Processing and Computational Linguistics, pp. 136–145. Mexico City (2002)

Banerjee, S., Pedersen, T.: Extended gloss overlaps as a measure of semantic relatedness. In: Proceedings of the 18th International Joint Conference on Artificial Intelligence, pp. 805–810. Acapulco, Mexico (2003)

Bruce, R., Wiebe, J., Pedersen, T.: The measure of a model. In: Proceedings of the Conference on Empirical Methods in Natural Language Processing, pp. 101–112. Philadelphia, PA (1996)

Collins, A.H., Quillian, M.R.: Retrieval time from semantic memory. J. Verb. Learn. Verb. Be. **8**, 240–247 (1969)

Fellbaum, C. (ed.): WordNet: An Electronic Lexical Database. The MIT Press, Cambridge (1998)

Hristea, F.: Recent advances concerning the usage of the Naïve Bayes model in unsupervised word sense disambiguation. Int. Rev. Comput. Softw. **4**(1), 58–67 (2009)

Hristea, F., Popescu, M., Dumitrescu, M.: Performing word sense disambiguation at the border between unsupervised and knowledge-based techniques. Artif. Intell. Rev. **30**(1), 67–86 (2008)

Hristea, F., Popescu, M.: Adjective sense disambiguation at the border between unsupervised and knowledge-based techniques. Fundam. Inf. **91**(3–4), 547–562 (2009)

Kay, M.: The concrete lexicon and the abstract dictionary. In: Proceedings of the Fifth Annual Conference of the UW Centre for the New Oxford English Dictionary, pp. 35–41 (1989)

Leacock, C., Towell, G., Voorhees, E.: Corpus-based statistical sense resolution. In: Proceedings of the ARPA Workshop on Human Language Technology, Princeton, pp. 260–265. New Jersey (1993)

Miller G.A.: Nouns in WordNet: a lexical inheritance system. Int. J. Lexicogr. **3**(4), 245–264 (1990)

Miller, G.A.: WordNet: a lexical database. Commun. ACM **38**(11), 39–41 (1995)

Miller, G.A.: Nouns in WordNet. In: Fellbaum, C. (ed.) WordNet: An Electronic Lexical Database, pp. 23–46. The MIT Press, Cambridge (1998)

Miller, G.A., Beckwith, R., Fellbaum, C., Gross, D., Miller, K.: WordNet: an on-line lexical database. J. Lexicogr. **3**(4), 234–244 (1990)

Miller, G.A., Hristea, F.: WordNet nouns: classes and instances. Comput. Linguist. **32**(1), 1–3 (2006)

Pedersen, T., Bruce, R.: Knowledge lean word-sense disambiguation. In: Proceedings of the 15th National Conference on Artificial Intelligence, pp. 800–805. Madison, Wisconsin (1998)

Chapter 4
Syntactic Dependency-Based Feature Selection

Abstract The feature selection method we are presenting in this chapter makes use of syntactic knowledge provided by dependency relations. Dependency-based feature selection for the Naïve Bayes model is examined and exemplified in the case of adjectives. Performing this type of knowledge-based feature selection places the disambiguation process at the border between unsupervised and knowledge-based techniques. The discussed type of feature selection and corresponding disambiguation method will once again prove that a basic, simple knowledge-lean disambiguation algorithm, hereby represented by the Naïve Bayes model, can perform quite well when provided knowledge in an appropriate way. Our main conclusion will be that the Naïve Bayes model reacts well in the presence of syntactic knowledge of this type and that dependency-based feature selection for the Naïve Bayes model is a reliable alternative to the WordNet-based semantic one.

Keywords Bayesian classification · Word sense disambiguation · Unsupervised disambiguation · Knowledge-based disambiguation · Dependency relations · Dependency-based feature selection

4.1 Introduction

The present chapter focuses on an entirely different way of performing feature selection, that is equally knowledge-based. The Naïve Bayes model will be fed knowledge of a totally different nature than the one examined in Chap. 3. With the benefits of placing the disambiguation process at the border between unsupervised and knowledge-based techniques having already become obvious (see Chap. 3), our next concern is to augment the role of linguistic knowledge in informing the construction of the semantic space for WSD based on an underlying Naïve Bayes model. This chapter investigates the usage of syntactic features provided by dependency relations, as defined by the classical Dependency Grammar formalism.[1] Although dependency-based semantic space models have been studied and discussed by several authors (Padó and Lapata 2007; Năstase 2008; Chen et al. 2009), to our knowledge, grammatical dependencies have been used in conjunction with the Naïve Bayes model only very recently (Hristea and Colhon 2012).

The semantic space proposed to the Naïve Bayes model for unsupervised WSD in (Hristea and Colhon 2012) is based on syntactic knowledge, more precisely on dependency relations, extracted from natural language texts via a syntactic parser. The resulting dependency relations will indicate those words (features) which should be part of the disambiguation vocabulary when trying to decrease the number of parameters that are to be estimated by the EM algorithm.[2] The corresponding disambiguation method, which we shall be presenting here, makes use (Hristea and Colhon 2012) of a PCFG parser, namely the Stanford parser (Klein and Manning 2003), in order to extract syntactic dependency relations that will indicate the disambiguation vocabulary required by the Naïve Bayes model.

Dependency relations are considered a linguistically rich representation where fixed word order is not required, argument structure differences can be captured, different types of contexts can be selected and words do not have to co-occur within a small, fixed context window (Padó and Lapata 2007). Such properties have recommended dependency relations as appropriate for feeding syntactic knowledge to the Naïve Bayes model.

The discussed disambiguation method introduces (Hristea and Colhon 2012) dependency-based feature selection in the case of adjectives and compares test results with those obtained when using the disambiguation vocabulary previously generated (see Chap. 3) by WordNet. Two totally different ways of feeding knowledge of different natures to a knowledge-lean algorithm with an underlying Naïve Bayes model are ultimately compared in the case of unsupervised WSD. The discussed method will once again prove that a basic, simple knowledge-lean disambiguation algorithm, hereby represented by the Naïve Bayes model, can perform quite well when provided knowledge in an appropriate way, a remark also made by (Ponzetto and Navigli 2010).

Although the discussed disambiguation method has so far been tested (Hristea and Colhon 2012) only in the case of adjectives, it is our belief that the obtained results (see Sect. 4.3.2) should initiate an open discussion concerning the *type of knowledge* that is best suited for the Naïve Bayes model when performing the task of unsupervised WSD.

---

[1] Dependency grammar (DG) is a class of syntactic theories developed by Lucien Tesnière (1959). Within this theory, syntactic structure is determined by the grammatical relations existing between a word (a *head*) and its *dependents*.

[2] See the mathematical model presented in Chap. 2.

4.2 A Dependency-Based Semantic Space for WSD with a Naïve Bayes Model

In the case of unsupervised WSD with an underlying Naïve Bayes model, our intention is to construct a semantic space that takes into account *syntactic relations*. The Naïve Bayes model will be fed syntactic knowledge based on the consideration that "because syntax-based models capture more linguistic structure than word-based models, they should at least in theory provide more informative representations of word meaning" (Padó and Lapata 2007). The choice of the syntactic formalism to be used is not an easy one. The main concern (Hristea and Colhon 2012) was that of having the semantic relationships between concepts and the words that lexicalize them mirrored in some way, considering that semantic knowledge of this type had already proven useful in the disambiguation process (Hristea et al. 2008; Hristea 2009; Hristea and Popescu 2009). Once again following the line of reasoning of Padó and Lapata (2007), "an ideal syntactic formalism should abstract over surface word order, mirror semantic relationships as closely as possible, and incorporate word-based information in addition to syntactic analysis ... These requirements point towards dependency grammar, which can be considered as an intermediate layer between surface syntax and semantics".

Despite the various existing linguistic theories, which lead to different ways of viewing sentence structure and therefore syntactic analysis, most linguists today agree that at the heart of sentence structure are *the relations among words*. These relations refer either to grammatical functions (subject, complement etc.) or to the links which bind words into larger units like phrases or even sentences. The dependency grammar approach to syntactic analysis takes into consideration the latter, viewing each word as *depending* on another word that links it to the rest of the sentence. Unlike generative grammars therefore, dependency grammars (DG) are not based on the notion of constituent but on the direct relations existing among words.

The relation between the *dependent* word and the word on which it depends (the *head*) is at the basis of DG. The syntactical analysis of a sentence signifies, from the point of view of DG, the description of all *dependency relations* (between the head and the dependent) which occur among all words of the sentence. Any word should depend exactly on one other word (the head), with the exception of the main predicate in the sentence which depends on no other word. Several words may depend on the same head. The dependency relations may or may not lack directionality[3] (from head to dependent) in the relation between words, according to which variant or alternative dependency-based grammatical theory[4] is used. A variety of dependency relations may exist among the words of a sentence if no restrictions are specified. The role of dependency grammars is mainly that of specifying the restrictions which the dependency relations should meet so that the structure they define is linguistically

---

[3] The relations between the dependent and the head are usually represented by an arch.

[4] See also *Link grammar* (Sleator and Temperley 1991, 1993) and *Word grammar* (Hudson 1984).

correct. The *dependency structure* will specify, in the case of each word, what other word it depends on. The dependencies indicated by the dependency structure of a sentence map straightforwardly onto a directed graph representation, in which the words of the represented sentence are nodes in the graph and grammatical relations are edge labels. It is various combinations of such dependencies[5] that will form the context over which we shall be constructing the semantic space for WSD. Just as Padó and Lapata (2007), we adopt the working hypothesis that syntactic structure in general and argument structure in particular are a close reflection of lexical meaning (Levin 1993). When using dependency relations we model meaning by quantifying the degree to which words occur in similar syntactic environments.

The Naïve Bayes model's reaction to knowledge of this type will be tested in the case of unsupervised adjective sense disambiguation.

4.2.1 Dependency-Based Feature Selection

Of the several existing dependencies parsers (Minipar, MaltParser, the Berkeley Parser, Stanford Parser, Link Grammar Parser) Hristea and Colhon (2012), whose line of reasoning we are following here, have chosen to use the Stanford Parser (Klein and Manning 2003) in order to automatically extract typed dependency parses of English sentences.

While the classical dependency-based linguistic theory does not allow the arches denoting the dependency relations to intersect (thus leading to an oriented graph which has no cycles), the dependency analysis performed by the Stanford parser can be either projective (disallowing crossing dependencies) or non-projective (permitting crossing dependencies). When using this specific syntactic parser the mentioned authors have performed a dependency syntactical analysis of non-projective type, in order to maximize the number of dependencies between content words. Although this may increase the number of features (words included into the disambiguation vocabulary) and therefore of parameters which must be estimated by the EM algorithm,[6] it was their belief that it could give a better indication of the ambiguous word's sense in context. The number of resulting features should then be decreased by taking into account only dependency relations of specific types (see Sect. 4.3.1).

Tests concerning the construction of the semantic space for WSD by feeding the Naïve Bayes model syntactic knowledge (provided by dependency relations) have so far concentrated (Hristea and Colhon 2012) on adjectives. However, we believe that the discussion which is to follow holds for all parts of speech (POS) and should only be subject to certain adaptations, depending on the particularities of each syntactic POS category. The mentioned authors' intention was to study the effectiveness of syntactic features (determined by dependency relations) as compared to semantic

---

[5] For which see Sect. 4.3.1.

[6] See the mathematical model presented in Chap. 2.

ones, more precisely to the ones provided by WordNet (see Chap. 3) which has been created in the spirit of understanding human semantic memory.

In order to inform the construction of the semantic space for WSD with syntactic knowledge of this type, Hristea and Colhon (2012) have conducted a two-stage experiment. At the first stage of their study, they have made no qualitative distinction between the different relations, by not taking into account the type of the involved dependencies.[7] This approach was inspired by existing syntax-based semantic space models, where the construction of the space is either based on all relations (Grefenstette 1994; Lin 1998) or on a fixed subset (Lee 1999), but always with no qualitative distinction between the different relations being made. At the second stage of their experiment Hristea and Colhon (2012) have taken into account the *dependency type*, thus informing the construction of the semantic space in a more linguistically rich manner. They have therefore eliminated certain paths (of the associated dependency graph) from the semantic space, on the basis of linguistic knowledge, by making use only of specific dependency relations, which are considered more informative than others relatively to the studied part of speech.

At both stages, their experiment takes place in the same type of setup. Namely, in defining the syntactic context of the target word, they have first taken into consideration *direct relationships*[8] between the target and other words (denoted by dependency relations where the target is either the head or a dependent and which correspond to paths of length 1 anchored[9] at the target in the associated dependency graph). They have subsequently considered *indirect relationships*[10] between the target and other words by taking into account paths of length 2 in the same associated dependency graph. At the present stage of their study Hristea and Colhon (2012) have limited their investigation to *second order dependencies*. However, the length (order) of these dependencies (paths in the associated graph) represent a parameter that can vary and which we consider analogous to the classical "window size" parameter. This parameter should have relatively small values, since it is a known fact that linguistically interesting paths are of limited length. By taking into account second order dependencies one additionally represents indirect semantic relations which could prove to be important.

When using dependency-based syntactic features the disambiguation vocabulary is formed by taking into account all words that participate in the considered dependencies.

In their experiments Hristea and Colhon (2012) have considered both dependencies having directionality and dependencies lacking it. Contrary to other studies (Padó and Lapata 2007), which consider that "directed paths would limit the context too severely", they have taken into account both undirected and directed paths, with

---

[7] They have only eliminated the potentially unuseful relations—for WSD—provided by the Stanford parser, such as: determiner, predeterminer, numeric determiner, punctuation relations, etc.

[8] In what follows, such dependencies will be called *first order dependencies*.

[9] A path anchored at the target word w is a path in the dependency graph starting at w. If the dependency relations have directionality, leading to an associated oriented graph, a path anchored at w is either a path starting at w or arriving at w.

[10] In what follows, such dependencies will be called *second order dependencies*.

the latter providing the best test results (see Sect. 4.3.2). As will be seen, the Naïve Bayes model seems to react well to the directionality of dependency relations.

4.3 Empirical Evaluation

In order to compare their disambiguation results to those of other previous studies (Hristea et al. 2008; Hristea and Popescu 2009) that had made use of the same Naïve Bayes model, trained with the EM algorithm, but had performed semantic WordNet-based feature selection, Hristea and Colhon (2012) try to disambiguate the same target adjectives using the same corpora. Specifically, they report disambiguation results in the case of adjectives *common* and *public* (see Sect. 4.3.2).

4.3.1 Design of the Experiments

With respect to adjectives Hristea and Colhon (2012) have used as test data the (Bruce et al. 1996) data containing twelve words taken from the ACL/DCI Wall Street Journal corpus and tagged with senses from the Longman Dictionary of Contemporary English. They have chosen this data set for their tests concerning adjectives since it has equally been used in the case of the Hristea et al. (2008) approach to WSD (WordNet-based feature selection), to which they were comparing the results of their own disambiguation method (dependency-based feature selection). As already mentioned, test results were reported in the case of two adjectives, *common* and *public*. The senses of *common* that have been taken into consideration and their frequency distribution are shown in Table 4.1, while Table 4.2 provides the same type of information corresponding to the adjective *public*. In these tables[11] *total count* represents the number of occurrences in the corpus of each word, with each of the adjectives being limited to the 3 most frequent senses, while *count* gives the percentage of occurrence corresponding to each of these senses.

In order for the experiments to be conducted, the data set was preprocessed (Hristea and Colhon 2012) in the usual required way for WSD: the stop words were eliminated and Porter stemmer was applied to the remaining words.

When performing syntactic (dependency-based) feature selection, at the first stage of the testing process, the mentioned authors have taken into account both directed and undirected dependency relations. No information concerning the types of the considered dependencies was used. At this stage of their study, they have designed a set of eight experiments (with the first two referring to undirected dependencies and the following six referring to directed ones).

---

[11] Which are the same as those showing the distribution of senses of *common* and *public*, respectively in Chap. 3.

Table 4.1 Distribution
of senses of *common*

Sense	Count
As in the phrase "common stock"	84 %
Belonging to or shared by 2 or more	8 %
Happening often; usual	8 %
Total count	1,060

Table 4.2 Distribution
of senses of *public*

Sense	Count
Concerning people in general	68 %
Concerning the government and people	19 %
Not secret or private	13 %
Total count	715

The first performed experiment (Hristea and Colhon 2012) considers all undirected first order dependencies anchored at the target word. All words participating in these dependencies (with the exception of the target) will be included in the so-called disambiguation vocabulary. This experiment is referred to in Table 4.3[12] of Sect. 4.3.2 (corresponding to adjective *common*) and Table 4.4[13] of Sect. 4.3.2 (corresponding to adjective *public*) as *Undirected first order dependencies*.

The second performed experiment (Hristea and Colhon 2012) also refers to undirected dependencies. It takes into account all first order and second order dependencies which are anchored at the target word. All words participating in these dependencies are part of the disambiguation vocabulary. This experiment is referred to in Table 4.3 of Sect. 4.3.2 (corresponding to adjective *common*) and Table 4.4 of Sect. 4.3.2 (corresponding to adjective *public*) as *Undirected first and second order dependencies*.

The following six experiments were all designed (Hristea and Colhon 2012) with reference to dependency relations that have directionality.[14] Both dependencies that view the target word as head[15] and dependencies that view it as dependent[16] have been considered. Within this group of experiments, the first two refer to directed first order dependencies and the following four to directed first and second order dependencies, respectively (see Tables 4.3 and 4.4 of Sect. 4.3.2).

The third performed experiment which is presented in Tables 4.3 and 4.4 of Sect. 4.3.2 views the target word as *head*. It takes into account all head-driven dependencies of first order anchored at the target word and collects all corresponding dependents, which form the considered disambiguation vocabulary. This experiment must be looked up, in Table 4.3 of Sect. 4.3.2 (corresponding to adjective *common*)

---

[12] Undertaken from (Hristea and Colhon 2012).

[13] Undertaken from (Hristea and Colhon 2012).

[14] The considered directionality is from head to dependent.

[15] In what follows, these dependencies will be called *head-driven dependencies*.

[16] In what follows, these dependencies will be called *dependent-driven dependencies*.

and Table 4.4 of Sect. 4.3.2 (corresponding to adjective *public*) under *Directed first order dependencies*. It is referred to as *Head-driven dependencies*.

The fourth performed experiment views the target word as *dependent*. It takes into account all dependent-driven dependencies of first order anchored at the target word and collects all corresponding heads, which form the considered disambiguation vocabulary. This experiment must also be looked up, in Table 4.3 of Sect. 4.3.2 (corresponding to adjective *common*) and Table 4.4 of Sect. 4.3.2 (corresponding to adjective *public*) under *Directed first order dependencies*. It is referred to as *Dependent-driven dependencies*.

The fifth performed experiment views the target word as *head*. It takes into account all first order head-driven dependencies anchored at the target word and collects all corresponding dependents. Furthermore, it takes into consideration all first order head-driven dependencies anchored at the previously obtained dependents and collects the corresponding dependents of these dependents.[17] All such collected words are included in the disambiguation vocabulary. This experiment must be looked up, in Table 4.3 of Sect. 4.3.2 (corresponding to adjective *common*) and Table 4.4 of Sect. 4.3.2 (corresponding to adjective *public*) under *Directed first and second order dependencies* and *Head-driven dependencies* respectively. It is referred to as *Two head-driven dependencies*.

The sixth performed experiment also views the target word as *head*. It takes into account all first order head-driven dependencies anchored at the target word and collects all corresponding dependents. Furthermore, it takes into consideration all first order dependent-driven dependencies anchored at the previously obtained dependents and collects the corresponding heads of these dependents.[18] All such collected words are included in the disambiguation vocabulary. This experiment must be looked up, in Table 4.3 of Sect. 4.3.2 (corresponding to adjective *common*) and Table 4.4 of Sect. 4.3.2 (corresponding to adjective *public*) under *Directed first and second order dependencies* and *Head-driven dependencies* respectively. It is referred to as *Head-driven dependencies and dependent-driven dependencies*.

The seventh performed experiment views the target word as *dependent*. It takes into account all first order dependent-driven dependencies anchored at the target word and collects all corresponding heads. Furthermore, it takes into consideration all first order dependent-driven dependencies anchored at the previously obtained heads and collects the corresponding heads of these heads.[19] All such collected words are included in the disambiguation vocabulary. This experiment must be looked up, in Table 4.3 of Sect. 4.3.2 (corresponding to adjective *common*) and Table 4.4 of

---

[17] The case *Two head-driven dependencies* can be summarized as follows: let us denote the target word by A; collect all words of type B and C such that B is a dependent of A and C is a dependent of B.

[18] The case *Head-driven dependencies and dependent-driven dependencies* can be summarized as follows: let us denote the target word by A; collect all words of type B and C such that B is a dependent of A and B is a dependent of C.

[19] The case *Two dependent-driven dependencies* can be summarized as follows: let us denote the target word by A; collect all words of type B and C such that A is a dependent of B and B is a dependent of C.

Sect. 4.3.2 (corresponding to adjective *public*) under *Directed first and second order dependencies* and *Dependent-driven dependencies* respectively. It is referred to as *Two dependent-driven dependencies*.

The eighth and last performed experiment also views the target word as *dependent*. It takes into account all first order dependent-driven dependencies anchored at the target word and collects all corresponding heads. Furthermore, it takes into consideration all first order head-driven dependencies anchored at the previously obtained heads and collects the resulted dependents of these heads.[20] All such collected words are included in the disambiguation vocabulary. This experiment must be looked up, in Table 4.3 of Sect. 4.3.2 (corresponding to adjective *common*) and Table 4.4 of Sect. 4.3.2 (corresponding to adjective *public*) under *Directed first and second order dependencies* and *Dependent-driven dependencies* respectively. It is referred to as *Dependent-driven dependencies and head-driven dependencies*.

Contrary to more general comments made in other studies (Padó and Lapata 2007), as far as WSD with an underlying Naïve Bayes model is concerned, test results (Hristea and Colhon 2012) will show (see Sect. 4.3.2) that considering directionality of the dependency relations is essential when forming the disambiguation vocabulary. The Naïve Bayes model will be shown to react well, as clustering technique, to the directionality of dependency relations.

During the second stage of their testing process, Hristea and Colhon (2012) have been taking into account the *type* of the dependency relations and have chosen only typed dependencies that have been considered relevant for the study of adjectives. Such typed dependencies were the providers of the words to be included in the disambiguation vocabulary.

Of the various dependency relations provided by the Stanford parser, Hristea and Colhon (2012) have chosen a restricted set of such relations (that they have viewed as minimal) in order to conduct their experiments. The chosen dependency relations (Hristea and Colhon 2012) are: *adjectival modifier, nominal subject, noun compound modifier* and *preposition collapsed*. The latter, which is not typical for adjectives, is a common Stanford collapsed dependency relation. In the considered non-projective analysis of the Stanford parser, the dependencies that involve prepositions, conjunctions or multi-word constructions are collapsed in order to get direct dependencies between content words (de Marneffe and Manning 2008). In this case, the relation is applicable for adjectival constructs where the adjective can be accompanied/intensified by an adverb particle such as "more common" or "very difficult". As a general rule in choosing the dependency type, however, the mentioned authors have constantly looked for, or looked at, the noun[21] that the target adjective modifies.

---

[20] The case *Dependent-driven dependencies and head-driven dependencies* can be summarized as follows: let us denote the target word by A; collect all words of type B and C such that A is a dependent of B and C is a dependent of B.

[21] This principle, which gives the nominal information priority, while the adjectival information is evaluated strictly within the range allowed by the nominal one, has guided Hristea and Colhon (2012) when choosing the *nominal subject* relation, for instance. This relation refers to the predicative form of the adjective linked via a copula verb to the noun that the adjective modifies.

In the case in which the head role of the target adjective is not imposed, that is in the case of undirected dependency relations, Hristea and Colhon (2012) also take into consideration the *adjectival modifier* relation, a very frequent dependency relation for adjectives that connects them as dependents directly to the noun they modify. This is probably the most informative of the considered relations and corresponds to the *attribute* semantic relation which has been used (Hristea et al. 2008; Hristea and Popescu 2009) when performing WordNet-based feature selection (see Chap. 3).

The experiments conducted by Hristea and Colhon (2012) have once again used both first order and second order dependencies. In order to obtain second order dependencies, the first order dependency relations were composed only with the modifier-type relations for nouns, as the first order relations (usually) return the modified noun of the target adjective (in this case by means of the previously specified relations—*nominal subject* and *preposition collapsed*). The modifier-type relations that were considered are *adjectival modifier* (returning the modifying adjective) and *noun compound modifier* (for the modifying noun).

The presented test results (see Sect. 4.3.2) correspond to experiments performed with this restricted set of chosen dependency relations, which is meant to ensure a minimal number of features for WSD, as well as a restricted number of parameters to be estimated by the EM algorithm.[22] However, experiments of the same type could be conducted with an enlarged set of such relations, a choice which should be made according mainly to linguistic criteria and by the linguistic community.

During this second stage of their study, which takes into account the type of the involved dependencies, Hristea and Colhon (2012) have designed a set of four experiments. Both directed and undirected dependency relations have again been considered.

The first performed experiment uses undirected first order dependencies. The considered relations are *adjectival modifier*, *preposition collapsed* and *nominal subject*, respectively. The disambiguation vocabulary is formed, as before, with all words participating in these dependency relations (with the exception of the target adjective). This experiment is referred to in Table 4.5[23] of Sect. 4.3.2 (corresponding to adjective *common*) and in Table 4.6[24] of Sect. 4.3.2 (corresponding to adjective *public*) as *Undirected first order dependencies*.

The second performed experiment also refers to undirected dependencies, this time of second order as well. The disambiguation vocabulary is formed with all words provided by the undirected first order dependencies of the previous experiment, to which all words indicated by the considered undirected second order dependencies are added. When forming the second order dependencies the following modifier-type dependency relations are used: *adjectival modifier* and *noun compound modifier*. This experiment is referred to in Table 4.5 of Sect. 4.3.2 (corresponding to adjective *common*) and in Table 4.6 of Sect. 4.3.2 (corresponding to adjective *public*) as *Undirected first and second order dependencies*.

---

[22] See the mathematical model presented in Chap. 2.

[23] Undertaken from (Hristea and Colhon 2012).

[24] Undertaken from (Hristea and Colhon 2012).

The next two experiments were both designed (Hristea and Colhon 2012) with reference to dependency relations that have directionality. Within both experiments the target adjective is viewed as *head*[25] only. Therefore all considered dependencies are head-driven ones.

The third performed experiment takes into account first order head-driven dependencies and forms the disambiguation vocabulary with all words they provide. The dependency relations that were used (Hristea and Colhon 2012) are *preposition collapsed* and *nominal subject*. All these dependency relations take the target adjective as the head word and, as consequence, the resulted disambiguation vocabulary is made of all the dependents of the target. This experiment is referred to in Table 4.5 of Sect. 4.3.2 (corresponding to adjective *common*) and in Table 4.6 of Sect. 4.3.2 (corresponding to adjective *public*) as *Head-driven first order dependencies*.

The fourth and last performed experiment refers to first and second order head-driven dependencies. The disambiguation vocabulary is formed with all words provided by the first order head-driven dependencies of the previous experiment, to which all words indicated by the considered second order head-driven dependencies, representing dependents of the target's dependents, are added. The experiment therefore considers the head role of both the target and of its dependents. When forming the second order dependencies, the following modifier-type dependency relations are used (Hristea and Colhon 2012): *adjectival modifier* and *noun compound modifier*. This experiment is referred to in Table 4.5 of Sect. 4.3.2 (corresponding to adjective *common*) and in Table 4.6 of Sect. 4.3.2 (corresponding to adjective *public*) as *Head-driven first and second order dependencies*.

Test results have shown (see Sect. 4.3.2) that taking into account the *type* of the dependency relations when forming the disambiguation vocabulary for the Naïve Bayes model is of the essence.

4.3.2 Test Results

Performance is evaluated in terms of accuracy, just as in Chap. 3. In the case of unsupervised disambiguation defining accuracy is not as straightforward as in the supervised case. Our objective is to divide the I given instances of the ambiguous word into a specified number K of sense groups, which are in no way connected to the sense tags existing in the corpus. In all performed experiments, sense tags are used only in the evaluation of the sense groups found by the unsupervised learning procedure. These sense groups must be mapped to sense tags in order to evaluate system performance. As in previous studies (Hristea et al. 2008; Hristea and Popescu 2009),

---

[25] This is the approach suggested by the first series of performed experiments, which had disregarded the dependency type. Test results have shown (see Sect. 4.3.2) that directionality of the relations counts and that the best disambiguation results are obtained when the target word plays the role of head.

Hristea and Colhon (2012) have used the mapping that results in the highest classification accuracy.[26]

Test results are presented in Tables[27] 4.3, 4.4, 4.5 and 4.6. Each result represents the average accuracy and standard deviation obtained by the learning procedure over 20 random trials while using a threshold ε having the value 10^{-9}.

Apart from accuracy, the following type of information is also included in Tables 4.3, 4.4, 4.5 and 4.6: number of features resulting in each experiment and percentage of instances having only null features (i.e. containing no relevant information).

As previously mentioned, within the present approach to disambiguation, the value of a feature is given by the number of occurrences of the corresponding word in the given context window. Since the process of feature selection is based on the restriction of the disambiguation vocabulary, it is possible for certain instances not to contain any of the relevant words forming this vocabulary. Such instances will have null values corresponding to all features. These instances do not contribute to the learning process. However, they have been taken into account in the evaluation stage of the presented experiments. Corresponding to these instances, the algorithm assigns the sense s_k for which the value $P(s_k)$ (estimated by the EM algorithm)[28] is maximal.

As far as the Stanford parser (Klein and Manning 2003) is concerned, the output was generated (Hristea and Colhon 2012) in dependency relation format (de Marneffe et al. 2006) and the data was preprocessed (Hristea and Colhon 2012) in the usual way: edges that do not connect open-class words were filtered out, words were lemmatized. The first of the mentioned operations could lead to some instances having only null features. However, corpus coverage will be much greater here than in the case of WordNet-based feature selection, where an independent knowledge source (WN) is used. Which makes the obtained results even more valuable.

Test results are presented in Tables 4.3 and 4.5 (corresponding to adjective *common*) and in Tables 4.4 and 4.6 (corresponding to adjective *public*).

At the first stage of the testing process, when no information concerning the dependency type was used (see the experiments designed in Sect. 4.3.1), the best obtained accuracy was 0.643 ± 0.09 in the case of adjective *common* and 0.607 ± 0.02 in the case of adjective *public*, respectively. These disambiguation results are more modest than the ones obtained in (Hristea et al. 2008) and in (Hristea and Popescu 2009) when performing WordNet-based feature selection (as described in Chap. 3). These studies report an accuracy of 0.775 ± 0.02 (obtained with 83 features and 19.2 % instances having only null features) in the case of adjective *common* and an accuracy of 0.559 ± 0.03 (obtained with 74 features and 43.3 % instances having only null features) in the case of adjective *public*, respectively.

---

[26] For more details concerning how to define accuracy in the case of unsupervised disambiguation, see Sect. 3.4.2 of Chap. 3.

[27] Undertaken from (Hristea and Colhon 2012).

[28] See the mathematical model presented in Chap. 2.

Table 4.3   First stage of the experiments corresponding to adjective *common*

Name of experiment			No. of features	Instances having only null features	Accuracy
Undirected first order dependencies			279	0.02	0.586 ± 0.09
Undirected first and second order dependencies			553	0.05	0.532 ± 0.09
Directed first order dependencies	*Head-driven dependencies*		135	0.00	0.616 ± 0.07
	Dependent-driven dependencies		178	0.02	0.547 ± 0.05
Directed first and second order dependencies	*Head-driven dependencies*	*Two head-driven dependencies*	224	0.01	0.643 ± 0.09
		Head-driven dependencies and dependent-driven dependencies	154	0.00	0.614 ± 0.10
	Dependent-driven dependencies	*Two dependent-driven dependencies*	281	0.03	0.517 ± 0.07
		Dependent-driven dependencies and head-driven dependencies	336	0.03	0.540 ± 0.10

When analyzing Tables 4.3 and 4.4 several conclusions have been drawn in (Hristea and Colhon 2012).

The best result is never obtained in the presence of undirected dependencies. When using undirected dependencies, taking into account second order dependencies as well does not increase accuracy in the case of adjective *common* and leads to only a slight increase in accuracy in the case of adjective *public*.

The highest accuracy is attained within the same testing setup, both corresponding to adjective *common* and corresponding to adjective *public*—namely in the case *Two head-driven dependencies*—which takes into consideration the head role of both the target and of its dependents.[29] Disambiguation accuracy obtained in this case is higher than the one attained by head-driven directed first order dependencies. A result which suggests that, when ignoring the dependency type, one should move to second order dependencies for increasing accuracy.

---

[29] Let us note that accuracy is always higher in the case *Two head-driven dependencies* than in the case *Head-driven dependencies and dependent-driven dependencies*, which shows that, in the case of directed first and second order dependencies, it is essential to consider the head role not only of the target word but also of its dependents.

Table 4.4 First stage of the experiments corresponding to adjective *public*

Name of experiment			No. of features	Instances having only null features	Accuracy
Undirected first order dependencies			340	0.03	0.424 ± 0.04
Undirected first and second order dependencies			698	0.07	0.436 ± 0.04
Directed first order dependencies	*Head-driven dependencies*		40	0.00	0.597 ± 0.03
	Dependent-driven dependencies		312	0.02	0.435 ± 0.02
Directed first and second order dependencies	*Head-driven dependencies*	*Two head-driven dependencies*	53	0.01	0.607 ± 0.02
		Head-driven dependencies and dependent-driven dependencies	51	0.01	0.569 ± 0.03
	Dependent-driven dependencies	*Two dependent-driven dependencies*	527	0.04	0.434 ± 0.04
		Dependent-driven dependencies and head-driven dependencies	515	0.05	0.426 ± 0.03

High accuracy is never attained in the case of dependent-driven dependencies. Considering solely the head role, corresponding to both first and second order dependencies, proves to be of the essence, a principle which has guided the second series of experiments. During which Hristea and Colhon (2012) have tried to decrease the number of features, and therefore of parameters that the EM algorithm must estimate,[30] by taking into account the *type* of the involved dependency relations (see the test results presented in Tables 4.5 and 4.6).

When comparing results of Table 4.3 with those of Table 4.5 and results of Table 4.4 with those of Table 4.6 respectively, it becomes obvious that disambiguation accuracy improves as a result of taking into consideration the type of the existing dependencies (Hristea and Colhon 2012).

When analyzing Tables 4.5 and 4.6, one notices that the best obtained accuracy was 0.775 ± 0.07 in the case of adjective *common* and 0.669 ± 0.01 in the case of adjective *public*, respectively. Both results were obtained with full corpus coverage, ensured by a small number of features (73 words corresponding to adjective *common* and 11 words corresponding to adjective *public*). Both accuracy values are superior

---

[30] See the mathematical model presented in Chap. 2.

Table 4.5 Second stage of the experiments corresponding to adjective *common*

Name of experiment	No. of features	Instances having only null features	Accuracy
Undirected first order dependencies	225	0.02	0.605 ± 0.09
Undirected first and second order dependencies	416	0.04	0.568 ± 0.09
Head-driven first order dependencies	73	0.00	0.775 ± 0.07
Head-driven first and second order dependencies	112	0.07	0.753 ± 0.08

Table 4.6 Second stage of the experiments corresponding to adjective *public*

Name of experiment	No. of features	Instances having only null features	Accuracy
Undirected first order dependencies	294	0.02	0.428 ± 0.03
Undirected first and second order dependencies	443	0.06	0.423 ± 0.03
Head-driven first order dependencies	11	0.00	0.669 ± 0.01
Head-driven first and second order dependencies	18	0.00	0.662 ± 0.01

to the ones obtained in the previous experiments, which did not take into account the dependency type, and are attained with a much smaller number of features. Maximal disambiguation accuracy corresponding to adjective *common* is practically the same as the one obtained when performing WordNet-based feature selection (see Chap. 3), the latter being attained with a more or less similar number of features but with less corpus coverage. In the case of adjective *public* accuracy increases significantly when performing dependency-based feature selection (as compared to WN-based feature selection), while the number of features used in disambiguation decreases significantly in spite of the ensured full corpus coverage. Both discussed accuracies were obtained (Hristea and Colhon 2012) within the same testing setup, namely in the case of head-driven first order dependencies, which take into account the head role of the target. In fact, this role is so significant that moving to second order dependencies becomes unnecessary in this case.[31]

Accuracies obtained when considering only undirected dependencies are always much lower, even when taking into account the dependency type. In both studied cases the head role of the target is of the essence.

---

[31] Disambiguation results are close but slightly inferior in the case of head-driven first and second order dependencies (see Tables 4.5 and 4.6).

The main significance of a dependency relation, that of indicating the head role of a word, proves itself to represent crucial information for the Naïve Bayes model when acting as clustering technique for unsupervised WSD.

4.3.2.1 Improving Accuracy in the Case of Dependency-Based Feature Selection for the Naïve Bayes Model

In order to study even further the reaction of the Naïve Bayes model to knowledge of syntactic type, specifically to dependency-based feature selection, and for the purpose of the present study, we have extended the two-stage experiment performed in (Hristea and Colhon 2012).

As already mentioned, Hristea and Colhon (2012) have performed a dependency syntactical analysis of non-projective type[32] in order to maximize the number of dependencies between content words, thus feeding the Naïve Bayes model as much knowledge of syntactic type as possible (relatively to the chosen set of dependency relations). Their best obtained results correspond to the experiments denoted *Head-driven first order dependencies* and *Head-driven first and second order dependencies* which are presented in Table 4.5 (corresponding to adjective *common*) and in Table 4.6 (corresponding to adjective *public*) respectively. For the purpose of the present discussion, we have repeated these two experiments, while performing a projective type[33] analysis when using the same Stanford parser. A smaller number of dependency relations will thus be taken into account, with a smaller number of words (features) being included into the disambiguation vocabulary. The Naïve Bayes model is fed less syntactic knowledge, with the number of parameters to be estimated by the EM algorithm[34] decreasing accordingly. From the minimal set of dependency relations considered in (Hristea and Colhon 2012) we have retained only those relations which ensure a projective type analysis (namely arches that do not cross, thus leading to an oriented graph which has no cycles). Specifically, for the experiment *Head-driven first order dependencies* we have considered only the *nominal subject* relation. In the case of the experiment *Head-driven first and second order dependencies* the relations *adjectival modifier* and *noun compound modifier* have been used. Test results are presented in Table 4.7 corresponding to adjective *common* and in Table 4.8 corresponding to adjective *public*, respectively.

As can be seen, disambiguation accuracy improves, compared to the one obtained in (Hristea and Colhon 2012) as a result of performing a projective type analysis, corresponding to both studied adjectives. In the case of adjective *common* accuracy increases from 77 % (non-projective) to 85 % (projective). In the case of adjective *public* the same accuracy increases from 66 % (non-projective) to 67 % (projective). In both cases the number of features used in disambiguation by the Naïve Bayes

---

[32] Which allows the arches denoting the dependency relations to intersect.

[33] Which does not allow the arches denoting the dependency relations to intersect, in accordance with the classical dependency linguistic theory.

[34] See the mathematical model presented in Chap. 2.

Table 4.7 Projective analysis corresponding to adjective *common*

Name of experiment	No. of features	Instances having only null features	Accuracy
Head-driven first order dependencies	28	0.00	0.858 ± 0.003
Head-driven first and second order dependencies	45	0.00	0.856 ± 0.01

Table 4.8 Projective analysis corresponding to adjective *public*

Name of experiment	No. of features	Instances having only null features	Accuracy
Head-driven first order dependencies	7	0.00	0.674 ± 0.007
Head-driven first and second order dependencies	12	0.00	0.674 ± 0.005

model decreases, with corpus coverage being fully ensured. This opens an entirely new line of investigation concerning the sensitivity of the Naïve Bayes model to syntactic features and to syntactic knowledge of this specific type. Both kinds of performed analyses show the Naïve Bayes model to react well to syntactic knowledge of dependency type.

4.4 Conclusions

This chapter has examined dependency-based feature selection and has tested the efficiency of such syntactic features in the case of adjectives. Performing this type of knowledge-based feature selection has once again placed the disambiguation process at the border between unsupervised and knowledge-based techniques, while reinforcing the benefits of combining the unsupervised approach to the WSD problem, based on the Naïve Bayes model, with usage of a knowledge source for feature selection. Specifically, syntactic knowledge provided by dependency relations, which has been under study here, seems to be even more useful to the Naïve Bayes model than the semantic one provided by WordNet (see Chap. 3).

Our main conclusion is that the Naïve Bayes model reacts well in the presence of syntactic knowledge of dependency type. The fact that 7 words (features) only, for instance, are sufficient in order to attain a higher disambiguation accuracy[35] than the one obtained by WordNet-based feature selection, while ensuring full corpus coverage, determines us to recommend syntactic dependency-based feature selection as a reliable alternative to the semantic one.

---

[35] In the case of adjective *public*; see Table 4.8 of Sect. 4.3.2.1.

The essence of the considered dependencies, namely the head role of a word (in our case the target), seems to represent crucial information for the Naïve Bayes model when acting as clustering technique for unsupervised WSD. Together with the dependency type, which should always be taken into account, since it expresses in what way the head links the dependent to the sentence in which they both occur. While directionality proves itself to be of the essence, *projective* syntactical analysis of dependency type should always be performed, which is in full accordance with the underlying linguistic theory.

As commented in Hristea and Colhon (2012), whether or not disambiguation accuracy can be improved by taking into consideration dependencies of various other types could represent a topic of discussion (primarily for the linguistic community). Whether or not this type of syntactic information can replace the mentioned semantic one should probably be subject to further investigation. And should also involve other parts of speech. Hristea and Colhon (2012) hope to have initiated an open discussion concerning the *type of knowledge* that is best suited for the Naïve Bayes model when performing the task of unsupervised WSD.

References

Bruce, R., Wiebe, J., Pedersen, T.: The Measure of a Model, CoRR, cmp-lg/9604018 (1996)

Chen, P., Bowes, C., Ding, W., Brown, D.: A fully unsupervised word sense disambiguation method using dependency knowledge. In: Human Language Technologies: The 2009 Annual Conference of the North American Chapter of the ACL, pp. 28–36 (2009)

de Marneffe, M.C., Manning, C.D.: Stanford typed dependencies manual. Technical Report, Stanford University (2008)

de Marneffe, M.C., MacCartney, B., Manning, C.D.: Generating typed dependency parses from phrase structure parses. In: Proceedings of LREC-06, pp. 449–454 (2006)

Grefenstette, G.: Explorations in Automatic Thesaurus Discovery. Kluwer Academic Publishers, Dordrecht (1994)

Hristea, F.: Recent advances concerning the usage of the Naïve Bayes model in unsupervised word sense disambiguation. Int. Rev. Comput. Softw. 4(1), 58–67 (2009)

Hristea, F., Colhon, M.: Feeding syntactic versus semantic knowledge to a knowledge-lean unsupervised word sense disambiguation algorithm with an underlying Naïve Bayes model. Fundam. Inform. 119(1), 61–86 (2012)

Hristea, F., Popescu, M.: Adjective sense disambiguation at the border between unsupervised and knowledge-based techniques. Fundam. Inform. 91(3–4), 547–562 (2009)

Hristea, F., Popescu, M., Dumitrescu, M.: Performing word sense disambiguation at the border between unsupervised and knowledge-based techniques. Artif. Intell. Rev. 30(1), 67–86 (2008)

Hudson, R.A.: Word Grammar. Blackwell, Oxford (1984)

Klein, D., Manning, C.D.: Accurate unlexicalized parsing. In: Proceedings of the 41st Meeting of the Association for Computational Linguistics (ACL 2003), pp. 423–430 (2003)

Lee, L.: Measures of distributional similarity. In: Proceedings of the 37th Annual Meeting of the Association for Computational Linguistics, pp. 25–32 (1999)

Levin, B.: English Verb Classes and Alternations: A Preliminary Investigation. University of Chicago Press, Chicago (1993)

Lin, D.: Automatic retrieval and clustering of similar words. In: Proceedings of the Joint Annual Meeting of the Association for Computational Linguistics and International Conference on Computational Linguistics, pp. 768–774 (1998)

Năstase, V.: Unsupervised all-words word sense disambiguation with grammatical dependencies. In: Proceedings of the Third International Joint Conference on Natural Language Processing, pp. 757–762 (2008)

Padó, S., Lapata, M.: Dependency-based construction of semantic space models. Comput. Linguist. **33**(2), 161–199 (2007)

Ponzetto, S.P., Navigli, R.: Knowledge-rich word sense disambiguation rivaling supervised systems. In: Proceedings of the 48th Annual Meeting of the Association for Computational Linguistics (ACL 2010), Uppsala, Sweden, ACL Press, pp. 1522–1531 (2010)

Sleator, D., Temperley, D.: Parsing English with a link grammar. Technical Report CMU-CS-91-196, Carnegie Mellon University, Pittsburgh, PA (1991)

Sleator, D., Temperley, D.: Parsing English with a link grammar. In: Proceedings of the Third International Workshop on Parsing Technologies (IWPT93), pp. 277–292 (1993)

Tesnière, L.: Eléments de syntaxe structurale. Klincksieck, Paris (1959)

Chapter 5
N-Gram Features for Unsupervised WSD with an Underlying Naïve Bayes Model

Abstract The feature selection method we are presenting in this chapter relies on web scale N-gram counts. It uses counts collected from the web in order to rank candidates. Features are thus created from unlabeled data, a strategy which is part of a growing trend in natural language processing. Disambiguation results obtained by web N-gram feature selection will be compared to those of previous approaches that equally rely on an underlying Naïve Bayes model but on completely different feature sets. Test results corresponding to the main parts of speech (nouns, adjectives, verbs) will show that web N-gram feature selection for the Naïve Bayes model is a reliable alternative to other existing approaches, provided that a "quality list" of features, adapted to the part of speech, is used.

Keywords Bayesian classification · Word sense disambiguation · Unsupervised disambiguation · Web-scale N-grams

5.1 Introduction

The present chapter focuses on an entirely different way of performing feature selection for the Naïve Bayes model, that relies on using web scale N-gram counts. The presented feature selection method was introduced in (Preoţiuc and Hristea 2012). To our knowledge, it represents a first attempt of using web N-gram features in unsupervised WSD in general, and in conjunction with the Naïve Bayes model as clustering technique for unsupervised WSD in particular. While creating features from unlabeled data, we are "helping" a simple, basic knowledge-lean disambiguation algorithm, hereby represented by the Naïve Bayes model, to significantly increase its accuracy as a result of receiving easily obtainable knowledge.

The proposed feature selection method (Preoţiuc and Hristea 2012) is based on the intuition that the most frequently occurring words near the target can give us a better indication of the sense which is activated than words being semantically similar that may not appear so often in the same context with the target word. The corresponding disambiguation method is unsupervised and knowledge-lean in the sense that it just requires the existence or the possibility to estimate N-gram counts for the target language corresponding to which the disambiguation process takes place. No information regarding the actual word senses will be used at any stage of the process. When using such features, the Naïve Bayes model will not require any sense definitions or sense inventories.

5.2 The Web as a Corpus

With respect to feature selection it is necessary to use those words that are the most relevant and distinctive for the target word. So, it is intuitive to think that these words are the ones that co-occur most often with the target. These words can be found by searching and performing an estimate over large corpora and the largest corpora available is the whole Web itself.

While the web provides an imense linguistic resource, collecting and processing data at web-scale is very timeconsuming. Previous research has relied on search engines to collect online information, but an alternative to this that has been developed more recently is to use the data provided in an N-gram corpus. An N-gram corpus is an efficient compression of large amounts of text as it states how often each sequence of words (up to length N) occurs.

The feature selection method that we are presenting here makes use of the Google Web 1T 5-gram Corpus Version 1.1, introduced in (Brants and Franz 2006), that contains English word N-grams (with N up to 5) and their observed frequency counts, calculated over 1 trillion words from the web and collected by Google in January 2006. The text was tokenized following the Penn Treebank tokenization, except that hyphenated words, dates, email addresses and URLs are kept as single tokens. The sentence boundaries are marked with two special tokens <S> and </S>. Words that occurred fewer than 200 times were replaced with the special token <UNK>. The data set has a N-gram frequency cutoff, that is N-grams that have a count that is less than 40 are discarded.

This corpus has been used in a variety of NLP tasks with good results. Yuret (2007) describes a WSD system that uses a statistical language model based on the Web 1T 5-gram dataset. The model is used to evaluate the likelihood of various substitutes for a word in a given context. These likelihoods are then used to determine the best sense for the word in novel contexts. (Bergsma et al. 2009) presents a unified view of using web-scale N-gram models for lexical disambiguation and uses the counts of 2–5 grams in a supervised method on the task of preposition selection, spelling correction or non-referential pronoun detection. In (Bergsma et al. 2010) web-scale N-gram data is used for supervised classification on a variety of NLP tasks such as: verb part-of-speech disambiguation, prenominal adjective ordering or noun compound bracketing. Islam and Inkpen (2009) have used the N-gram data for spelling correction, while Chang and Clark (2010) have made use of this data to check the acceptability of paraphrases in context.

Web-scale N-gram counts are used for the first time in unsupervised word sense disambiguation, as a mean of feature selection for the Naïve Bayes model, in (Preoţiuc and Hristea 2012).

In order to find the most frequent words that co-occur with the target word within a distance of N−1 words, one must take into consideration the N-grams in which the target word occurs. Thus, we can build different feature sets depending on the size of N and on the number of words to include in the feature set. These sets will be referred using the following convention: **n-w-t** represents the set containing the top **t** words occurring in **n**-grams together with the word **w**.

For example, *5-line-100* is the set constituted by the most frequent 100 (stemmed) words that co-occur in the Web with the word *line* within a distance of, at most, 4 words.

In order to build the feature set corresponding to the top t words occurring in N-grams of size n with the target word w, (n-w-t), Preoţiuc and Hristea (2012) have used the following processing directions:

- they have lowercased every occurrence in the N-gram corpus and have combined the counts for identical matches;
- for every number $k(k < n)$, they have built a list of words and counts, each representing word counts occurring at a distance of exactly k on each side of the target word;
- they have merged the counts from all $n - 1$ lists to get a complete list of words and counts that co-occur in a context window of size $n - 1$ with the target word w;
- they have removed the numbers, the punctuation marks, the special tokens (eg. <s>, <unk>), the words starting with special characters or symbols and the stopwords from the list;
- they have performed stemming using the Porter Stemmer on each feature set, merging counts for similar words whenever the case;
- they have sorted the word and counts pairs in descending order of their counts and have extracted the top t words.

Let us note the fact that, while in the context window only content words exist, within the N-grams stopwords may also occur. So it is not guaranteed that the N-grams show the counts of words appearing in a context window of N−1. Preoţiuc and Hristea (2012) have chosen to eliminate stopwords because they appear much too often in the corpora and, by using them as features, the model tends to put too much weight on these, as opposed to the content words that are the ones indicative of the word sense.

Despite the fact that the target words and the dataset we refer to in the experiments are in English, the feature selection method we are discussing here is language independent and can be applied with no extra costs to other languages for which we know or can estimate N-gram counts from large data. Recently, Google has released Web 1T 5-gram, 10 European Languages Version 1 (Brants and Franz 2009) consisting of word N-grams and their observed frequency counts for other ten European languages: Czech, Dutch, French, German, Italian, Polish, Portuguese, Romanian, Spanish and Swedish. The N-grams were extracted from publicly accessible web pages from October 2008 to December 2008 using the same conventions as for the English data set, with only the data being approximately 10 times smaller. Thus, the presented method can be used with no changes whatsoever to extract features for performing sense disambiguation corresponding to these languages as well.

Using a Web scale N-gram corpus implies performing counts that take into account all the possible senses of the target word. Automatically, when computing these counts, high frequency senses will have more words indicative of those senses than low frequency senses have. If the disambiguation setting is restricted to a specific domain (eg. medicine), the discussed method of feature extraction could be used with a N-gram corpus derived from large corpora of texts in that domain.

5.3 Experimental Results

Preoţiuc and Hristea (2012) have tested their proposed feature sets for the three main parts of speech: nouns, adjectives and verbs. They have drawn conclusions, that we shall be presenting here, with regard to each of these parts of speech.

5.3.1 Corpora

In order to compare their results with those of other previous studies (Pedersen and Bruce 1998; Hristea et al. 2008; Hristea 2009; Hristea and Popescu 2009) that have presented the same Naïve Bayes model, trained with the EM algorithm, but using other methods of feature selection, Preoţiuc and Hristea (2012) try to disambiguate the same target words using the same corpora.

In the case of nouns they have used as test data the *line* corpus (Leacock et al. 1993). This corpus contains around 4,000 examples of the word *line* (noun) sense-tagged with one of the 6 possible WordNet 1.5 senses. Examples are drawn from the WSJ corpus, the American Printing House for the Blind, and the San Jose Mercury. The description of the senses and their frequency distribution[1] are shown in Table 5.1.

In (Pedersen and Bruce 1998; Hristea et al. 2008) tests are also performed for only 3 senses of *line*. Preoţiuc and Hristea (2012) do not perform this comparison as their method is not relying on sense inventories. Therefore it is not possible to distinguish and take out the words that co-occur with the specific senses represented in the test set.

In the case of adjectives and verbs the mentioned authors have used as test data the corpus introduced in (Bruce et al. 1996) that contains twelve words taken from the ACL/DCI Wall Street Journal corpus and tagged with senses from the Longman Dictionary of Contemporary English.

---

[1] Which are the same as those considered in Chap. 3.

Table 5.1 Distribution
of senses of *line*

Sense	Count	Pct. (%)
Product	2,218	53,47
Written or spoken text	405	9,76
Telephone connection	429	10,34
Formation of people or things; queue	349	8,41
An artificial division; boundary	376	9,06
A thin, flexible object; cord	371	8,94
Total count	4,148	100

Table 5.2 Distribution
of senses of *common*

Sense	Count	Pct. (%)
As in the phrase "common stock"	892	84
Belonging to or shared by 2 or more	88	8
Happening often; usual	80	8
Total count	1,060	100

Table 5.3 Distribution
of senses of *public*

Sense	Count	Pct. (%)
Concerning people in general	440	68
Concerning the government and people	129	19
Not secret or private	90	13
Total count	659	100

Tests have been conducted for two adjectives, *common* and *public*, the latter being the one corresponding to which Pedersen and Bruce (1998) obtain the worst disambiguation results.

The senses of *common* and *public* that have been taken into consideration and their frequency distribution[2] are shown in Table 5.2 and in Table 5.3, respectively. In order to compare their results to those of (Pedersen and Bruce 1998; Hristea et al. 2008; Hristea and Popescu 2009), Preoţiuc and Hristea (2012) have also taken into account only the 3 most frequent senses of each adjective, as was the case in those studies.

For verbs, the part of speech which is known as being the most difficult to disambiguate, Preoţiuc and Hristea (2012) have performed tests corresponding to the verb *help* while considering the most frequent two senses of this word. The definition of the senses and the frequency distribution[3] are presented in Table 5.4.

In order for the experiments to be conducted, the data set was preprocessed (Preoţiuc and Hristea 2012) in the usual way: the stopwords, words with special characters and numbers were eliminated and stemming was applied to all remaining words, using the same Porter Stemmer as in the case of stemming the lists of feature words.

---

[2] Which are the same as those considered in Chaps. 3 and 4.

[3] Which are the same as those considered in Chap. 3.

Table 5.4 Distribution of senses of *help*

Sense	Count	Pct. (%)
To enhance-inanimate object	990	78
To assist-human object	279	22
Total count	1,269	100

5.3.2 Tests

As was the case in the mentioned previous studies that examine unsupervised WSD with an underlying Naïve Bayes model, studies to the results of which they are comparing their own disambiguation results, Preoţiuc and Hristea (2012) also evaluate performance in terms of accuracy. As it is well known, in the case of unsupervised disambiguation defining accuracy is not as straightforward as in the supervised case. The objective is to divide the I given instances of the ambiguous word into a specified number K of sense groups, which are in no way connected to the sense tags existing in the corpus. In the experiments, sense tags are used only in the evaluation of the sense groups found by the unsupervised learning method. These sense groups must be mapped to sense tags in order to evaluate system performance. As in the previously mentioned studies, in order to enable comparison, Preoţiuc and Hristea (2012) have used the mapping that results in the highest classification accuracy.

In the case when none of the words belonging to the feature set are found in the context window of the target, as in (Hristea et al. 2008; Hristea 2009; Hristea and Popescu 2009), the disambiguation method presented by Preoţiuc and Hristea (2012) assigns the instance to the cluster that has the greatest number of assignments. If the target word has a dominant sense, which is the case with all the considered test target words, lower coverage will determine an increase in the performance of the method when results are below the most frequent sense baseline (a very high one in the case of unsupervised WSD using the same underlying mathematical model). With respect to this, Preoţiuc and Hristea (2012) also define coverage as the percentage of instances in which at least one feature word occurs in the context window and, so, the assignment is performed by the Naïve Bayes classifier as opposed to a most frequent sense one.

Preoţiuc and Hristea (2012) show results that couple accuracy with coverage. They use a context window with varying size around the target word, the coverage for a feature set increasing accordingly with the enlargement of the window size.

As in (Hristea et al. 2008) each presented result represents the average accuracy obtained by the disambiguation method over 20 random trials while using a fixed threshold ε having the value 10^{-9}.

In what follows, we show the most significant test results that were obtained (Preoţiuc and Hristea 2012) in the case of all main parts of speech.

Within the graphs, the (Preoţiuc and Hristea 2012) results are designated by solid lines with different markers indicating the various parameters (n or t) that were used. The context window sizes vary and are listed in the corresponding text for

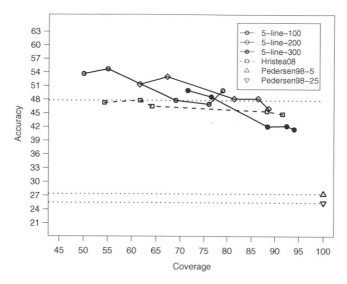

Fig. 5.1 Results for feature sets *5-line*

each part of speech. The Hristea et al. (2008) method is presented with a dashed line and always uses a context window of size 25. The variation in coverage is due to the different type of WordNet relations that were used, resulting in a different number of feature words. The results of the Pedersen and Bruce (1998) method are presented as well. We notice that here we always have just one value, corresponding to a 100 % coverage and to a size of 5 or 25 of the context window. This is due to the fact that the method of feature selection takes into consideration all the words in the vocabulary. Therefore, in this case, there are no contexts with no features. In each graph, corresponding to each of the other two previous methods, and in order to allow an easier visual comparison, Preoţiuc and Hristea (2012) have drawn a dotted black line to illustrate the highest accuracy obtained for that word by the respective method.

5.3.2.1 Test Results Concerning Nouns

In the case of the noun *line* results are presented in Fig. 5.1.[4]

The best results were obtained by using the most frequent words appearing in 5-gram with *line*, although results with a lower n were only slightly worse, as reported in (Preoţiuc and Hristea 2012).

Test results are presented (Preoţiuc and Hristea 2012) for context windows of size 4, 5, 10, 15 and 25 corresponding to each feature set. We observe the largest difference in favour of the Preoţiuc and Hristea feature selection method as resulting

---

[4] Reprinted here from (Preoţiuc and Hristea 2012).

in an accuracy of 54.7% (for context window 5 and feature set *5-line-100*) as compared to 47.8% for a similar coverage in (Hristea et al. 2008). For the feature sets *5-line-100* and *5-line-200*, the tests concerning web N-gram feature selection show better performances than any of the results of Hristea et al. (2008) and better, by a wide margin, than those of Pedersen and Bruce (1998). For some experiments, the method outperforms the most frequent sense baseline which, in this case, is situated at 53.47%.

The graph also shows that by increasing too much the number of features (*5-line-300*), the performance of the system decreases. This performance decreases even more when considering even larger feature sets ($t = 500$ or 1000—not shown on the graph for clarity).

We observe that when web N-gram feature selection is performed in the case of noun disambiguation, increasing the size of the context window (thus bringing more features into the process) does not bring improvements to the disambiguation results (taking into consideration the coverage-accuracy trade-off), as stated in other studies. As reported in (Preoţiuc and Hristea 2012), another interesting aspect is that, by every step in extending the context window, the coverage increases significantly. This remark is not valid, as we shall see, in the case of adjectives and verbs.

The obtained results (Preoţiuc and Hristea 2012) confirm the intuition that, in order to disambiguate a noun, the information in a wide context is useful and can contribute to the disambiguation process. Features taken from wider contexts are also good indicators for disambiguation.

5.3.2.2 Test Results Concerning Adjectives

With respect to adjectives, Preoţiuc and Hristea (2012) have considered the disambiguation of the polysemous words *common* and *public*. Test results are shown in Figs. 5.2[5] and 5.3,[6] respectively.

The best results were achieved by using the most frequent words appearing in bigrams with *common* and in 3-grams with *public* (although results with bigrams for *public* were close in terms of accuracy).

In the case of adjective *common* the results are presented for context windows of size 1, 2, 3, 4, 5 and 10. We observe the largest difference in favour of the Preoţiuc and Hristea (2012) feature selection method as resulting in an accuracy of 87.0%, as compared to 77.5%, the best result obtained in (Hristea et al. 2008). Again, almost all scores (16 out of 18 shown) are higher than the ones of the Hristea et al. (2008) method, with almost half of them exceeding the most frequent sense baseline (set at 84.0% in this case).

Corresponding to the adjective *public* test results are presented for context windows of size 2, 3, 4, 5 and 10. The Preoţiuc and Hristea (2012) best result is 58.7%

---

[5] Reprinted here from (Preoţiuc and Hristea 2012).

[6] Reprinted here from (Preoţiuc and Hristea 2012).

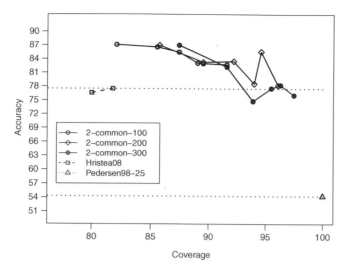

Fig. 5.2 Results for feature sets *2-common*

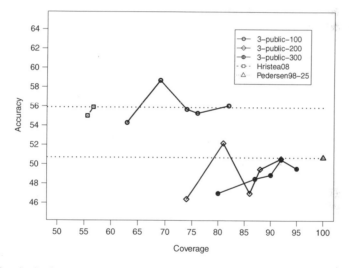

Fig. 5.3 Results for feature sets *3-public*

accuracy as compared to 55.9 % obtained with much smaller coverage in (Hristea et al. 2008).

We must keep in mind that, as we move to the right of the graph (increasing coverage), the results are more significant, because the bias of choosing the most frequent sense baseline for contexts with no features is reduced, due to the fact that the baseline has a very high value (84 and 68 % respectively).

For both adjectives, we observe that just by taking the most frequent 100 words in bigrams or trigrams and a very narrow context window (starting with size 1) we already

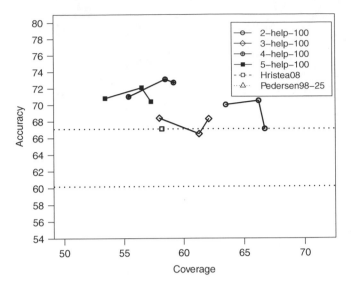

Fig. 5.4 Results for feature sets *help-100*

obtain a very high coverage, that increases at a low rate together with the enlargement of the context window. This corresponds to the linguistic argument that an adjective will appear together with the word it modifies, the latter representing the most frequent and important attribute when disambiguating the respective adjective. Results with wider N-grams were inferior by a distinctive margin (Preoţiuc and Hristea 2012).

5.3.2.3 Test Results Concerning Verbs

Corresponding to the verb *help* test results are shown in Fig. 5.4.[7]

As commented in (Preoţiuc and Hristea 2012), interestingly enough, the best results were achieved by using the top 100 words regardless of the order of the N-grams. The (Preoţiuc and Hristea 2012) top result was 73.1 % when using words from 4-grams and a context window of size 15, as compared to a maximum of 67.1 % in (Hristea et al. 2008), obtained with similar coverage. Out of 12 results, 11 were better than those in (Hristea et al. 2008), confirming the reliability of disambiguating using web N-gram feature sets.

Test results are presented for context windows of size 10, 15 and 25 respectively, as coverage is too low corresponding to smaller context windows. One can notice that coverage for this verb is very low compared to the case of the studied nouns and adjectives and that it increases by a very low margin with the enlargement of the context window.

---

[7] Reprinted here from (Preoţiuc and Hristea 2012).

This is also very linguistically intuitive because verbs usually appear in very different contexts. This makes feature selection more difficult and is the main reason why most studies conclude that this is the hardest to disambiguate part-of-speech.

As we are shown from the Preoţiuc and Hristea (2012) results, corresponding to all parts of speech, we can restate the fact that, by taking more, less related words (increasing t), the accuracy drops, a fact which emphasizes the need for a "quality list of features". The presented feature selection method (Preoţiuc and Hristea 2012) obtains very high results compared to Pedersen and Bruce (1998) in all tests, good results compared to Hristea et al. (2008) and sometimes exceeds the most frequent sense baseline, which is a high baseline to achieve using the Naïve Bayes model.

5.3.3 Adding Knowledge from an External Knowledge Source

While noting that web N-gram feature selection has provided the best disambiguation results so far, we are now trying to "help" the Naïve Bayes model, when acting as clustering technique for unsupervised WSD, by combining the described features with other, additional ones, coming from an external knowledge source. For the purpose of the present discussion, the chosen knowledge source will be WordNet.

Disambiguation results provided by WN-based feature selection are shown and commented in Chap. 3 corresponding to all major parts of speech (nouns, adjectives, verbs). WN-based feature selection has provided more modest disambiguation accuracies than those obtained when using web N-gram features. It is therefore natural to hope for an increase in accuracy when combining the WN-based features with those that have led to the best disambiguation results. In order to test this assumption we have performed[8] a great number of experiments that combine WN-based and web N-gram features. For enabling comparison, we have attempted to disambiguate the same polysemous words that have been discussed so far: the noun *line*, the adjectives *common* and *public* and the verb *help*. The same corpora have been used corresponding to each of these polysemous words.

In the case of the noun *line* we have designed experiments which perform discrimination between the 6 senses listed in Table 5.1. We have started by combining the two sets of features which had provided the best disambiguation results for each of the considered feature selection methods. In the case of WN-based feature selection this is the disambiguation vocabulary formed with WN synonyms, content words of the associated synset glosses and example strings, and all nouns coming from all hyponym and meronym synsets (see Chap. 3). This disambiguation vocabulary had brought an accuracy of 47.8 % (see Sect. 3.4.2.1). In the case of web N-gram feature selection the best disambiguation accuracy (54.7 %) has been obtained with the feature set *5-line-100* (see Sect. 5.3.2.1). When combining these two feature sets accuracy drops to 43.0 % (obtained with 70.3 % corpus coverage). Numerous

---

[8] Together with Daniel Preoţiuc.

other tests have been performed, none of which have led to the improvement of the disambiguation accuracy. Our best result is represented by an accuracy of 48.7 % (obtained with 95.8 % corpus coverage). As far as WN-based feature selection is concerned, this best result is obtained when considering the disambiguation vocabulary formed with all WN-synonyms and content words of the associated synset glosses and example strings, all nouns of hyponym synsets plus all content words of the associated glosses and example strings, as well as all nouns coming from the meronym synsets, to which all content words of the corresponding glosses and example strings are added. As far as web N-gram feature selection is concerned, the best obtained accuracy resulted when using the feature set *5-line-200*.

This best obtained accuracy (48.7 %) slightly improves the one resulting as best when performing WN-based feature selection alone, and does not come close to the best one obtained with web N-gram feature selection. In the case of nouns, the Naïve Bayes model does not react well to the combination of web N-gram features and WN-based ones.

In the case of the adjective *common* we have designed experiments which perform discrimination again between the 3 senses listed in Table 5.2. Our best obtained result is an accuracy of 87.2 % (with corpus coverage 83.5 %). This is very close to the obtained web N-gram result (87.0 %) and significantly improves the best obtained WN result (77.5 %). As far as feature sets are concerned, it is obtained corresponding to the extended WN vocabulary (all relations) discussed in Chap. 3, but leaving out antonyms, and to the web N-gram feature set *2-common-100*.

In the case of the adjective *public* we have designed experiments which perform discrimination between the 3 senses listed in Table 5.3. Out best obtained result is an accuracy of 56.4 % (with corpus coverage 73.2 %). This is lower than the obtained web N-gram result (58.7 %) and very slightly improves the best obtained WN result (55.9 %). As far as feature sets are concerned, it is obtained corresponding to the same extended WN vocabulary (all relations, including antonymy) discussed in Chap. 3 and to the web N-gram feature set *3-public-100*.

In the case of the verb *help* we have designed experiments which perform discrimination between the 2 senses listed in Table 5.4. Our best obtained result is an accuracy of 70.3 % (with corpus coverage 61.8 %). This is lower than the obtained web N-gram result (73.1 %) and improves the best obtained WN result (67.1 %). As far as feature sets are concerned, it is obtained corresponding to the extended WN vocabulary (all relations) discussed in Chap. 3 and to the web N-gram feature set *3-help-100*.

Our conclusion is that it is not worth combining these features of totally different natures, but it is recommendable to rather use web N-gram features alone.

5.4 Conclusions

This chapter has examined web N-gram feature selection for unsupervised word sense disambiguation with an underlying Naïve Bayes model.

The disambiguation method using N-gram features that we have presented here is unsupervised and uses counts collected from the web in a simple way, in order to rank candidates. It creates features from unlabeled data, a strategy which is part of a growing trend in natural language processing, together with exploiting the vast amount of data on the web. Thus, the method does not rely on sense definitions or inventories. It is knowledge-lean in the sense that it just requires the existence or the possibility to estimate N-gram counts for the target language corresponding to which the disambiguation process takes place. No information regarding the actual word senses is used at any stage of the process.

Comparisons have been performed with previous approaches that rely on completely different feature sets. In the case of all studied parts of speech, test results were better, by a wide margin, than those obtained when using local-type features (Pedersen and Bruce 1998). They have also indicated a superior alternative to WordNet feature selection for the Naïve Bayes model (see Chap. 3). Strictly as far as adjectives are concerned, results are more or less similar to those obtained when feeding the Naïve Bayes model syntactic knowledge of the studied type (see Chap. 4). Web N-gram feature selection seems a reliable alternative to projective dependency-based feature selection as well.

The experiments conducted for all three major parts of speech (nouns, adjectives, verbs) have provided very different results, depending on the feature sets that were used. These results are in agreement with the linguistic intuitions and indicate the necessity of taking into consideration feature sets that are adapted to the part of speech which is to be disambiguated.

Another conclusion we have come to, in the present study, is that, when using the Naïve Bayes model as clustering technique for unsupervised WSD, it is not recommended to combine features created from unlabeled data with those coming from an external knowledge source (such as WordNet).

Last but not least, the presented method has once again proven that a basic, simple knowledge-lean disambiguation algorithm, hereby represented by the Naïve Bayes model, can perform quite well when provided knowledge in an appropriate way.

References

Bergsma, S., Lin, D., Goebel, R.: Web-scale N-gram models for lexical disambiguation. In: Proceedings of the 21st International Joint Conference on Artificial Intelligence, pp. 1507–1512. Pasadena, California (2009)

Bergsma, S., Pitler, E., Lin, D.: Creating robust supervised classifiers via web-scale N-gram data. In: Proceedings of the 48th Annual Meeting of the Association for Computational Linguistics (ACL '10), pp. 865–874. Uppsala, Sweden (2010)

Brants, T., Franz, A.: Web 1T 5-gram corpus version 1.1. Technical Report, Google Research (2006)

Brants, T., Franz, A.: Web 1T 5-gram, 10 European languages version 1. Technical Report, Linguistic Data Consortium, Philadelphia (2009)

Bruce, R., Wiebe, J., Pedersen, T.: The Measure of a Model, CoRR, cmp-lg/9604018 (1996)

Chang, C.Y., Clark, S.: Linguistic steganography using automatically generated paraphrases. In: Human Language Technologies: The Annual Conference of the North American Chapter of the

Association for Computational Linguistics (HLT '10), pp. 591–599. Los Angeles, California (2010)

Hristea, F.: Recent advances concerning the usage of the Naïve Bayes model in unsupervised word sense disambiguation. Int. Rev. Comput. Softw. **4**(1), 58–67 (2009)

Hristea, F., Popescu, M., Dumitrescu, M.: Performing word sense disambiguation at the border between unsupervised and knowledge-based techniques. Artif. Intell. Rev. **30**(1), 67–86 (2008)

Hristea, F., Popescu, M.: Adjective sense disambiguation at the border between unsupervised and knowledge-based techniques. Fundam. Inform. **91**(3–4), 547–562 (2009)

Islam, A., Inkpen, D.: Real-word spelling correction using Google Web IT 3-grams. In: Proceedings of the Conference on Empirical Methods in Natural Language Processing (EMNLP '09), pp. 1241–1249. Singapore (2009)

Leacock, C., Towell, G., Voorhees, E.: Corpus-based statistical sense resolution. In: Proceedings of the ARPA Workshop on Human Language Technology, pp. 260–265. Princeton, New Jersey (1993)

Pedersen, T., Bruce, R.: Knowledge lean word-sense disambiguation. In: Proceedings of the 15th National Conference on Artificial Intelligence, pp. 800–805. Madison, Wisconsin (1998)

Preoţiuc-Pietro, D., Hristea, F.: Unsupervised word sense disambiguation with N-gram features. Artif. Intell. Rev. doi:10.1007/s10462-011-9306-y (2012)

Yuret, D.: KU: Word sense disambiguation by substitution. In: Proceedings of the 4th International Workshop on Semantic Evaluations (SemEval '07), pp. 207–214. Prague (2007)

Index

A
Accuracy, 27, 29, 31, 32, 45, 47, 49

B
Bag of words model, 10
Bayes classifier, 4, 9, 11
Bayesian classification, 9

C
Context window, 3, 10, 31, 32, 60
Contextual features, 10
Coverage, 60

D
Dependency grammar, 37
Dependency relation, 36–38, 40, 43–45
Dependency structure, 38
Dependency type, 39, 43, 52
Dependency-based feature selection, 36, 51
Dependency-based syntactic features, 39
Dependent-driven dependencies, 41
Direct relationship, 39
Disambiguation vocabulary, 3, 20, 23, 25, 28, 29, 31, 32, 36, 38, 39, 41–46, 50, 65

E
Expectation–Maximization (EM) algorithm, The, 12, 13, 15, 21, 27, 29, 30

F
Feature, 46
Feature selection, 3, 5, 29, 30, 32, 46, 61

F
First order dependencies, 39, 44
First order relations, 44

G
Gloss, 18, 19, 25, 28, 31

H
Head-driven
    dependencies, 41, 45

I
Indirect relationship, 39

K
Knowledge source, 2–4, 19, 32, 33
Knowledge-based
    disambiguation, 2, 4, 28
Knowledge-based feature
    selection, 3, 19, 51
Knowledge-lean disambiguation
    algorithm, 3, 36, 67
Knowledge-lean methods, 2, 4, 8

L
Line corpus, 22, 28, 30
Local features, 6, 21
Local-context features, 7
Local-type features, 8, 33, 67

M
Modifier-type relations, 44

F. T. Hristea, *The Naïve Bayes Model for Unsupervised Word Sense Disambiguation*,
SpringerBriefs in Statistics, DOI: 10.1007/978-3-642-33693-5, © The Author(s) 2013